THE
NATURE OF LIFE

XIIIth Nobel Conference

THE NATURE OF LIFE

edited by

William H. Heidcamp, Ph.D.

Chairman, Department of Biology
Gustavus Adolphus College
St. Peter, Minnesota

With contributions by **René Dubos,**
Sidney W. Fox, Bernard M. Loomer,
Peter Marler, and **Max Delbrück**

University Park Press
Baltimore

UNIVERSITY PARK PRESS
International Publishers in Science and Medicine
233 East Redwood Street
Baltimore, Maryland 21202

Typeset by Action Comp Co., Inc.
Manufactured in the United States of America by Universal Lithographers,
Inc., and The Optic Bindery Incorporated.

Proceedings of the XIIIth Nobel Conference, "The Nature of Life,"
held at Gustavus Adolphus College, St. Peter, Minnesota,
October 4–5, 1977.

Library of Congress Cataloging in Publication Data

Nobel Conference, 13th, Gustavus Adolphus College,
   1977.
   The nature of life.

   "Proceedings of the XIIIth Nobel Conference . . .
held at Gustavus Adolphus College, St. Peter, Minne-
sota, October 4–5, 1977."
   Includes index.
   1. Life (Biology)—Congresses.   I. Heidcamp,
William H., 1944–   II. Gustavus Adolphus
College, St. Peter, Minn.   III. Title.
QH501.N63   1977   577   78-9316
ISBN 0-8391-1280-7

Contents

Contributors

Max Delbrück, Ph.D.
California Institute of Technology
Division of Biology
Pasadena, California 91125

René Dubos, Ph.D., M.D.
The Rockefeller University
1230 York Avenue
New York, New York 10021

Sidney W. Fox, Ph.D.
Institute for Molecular and Cellular Evolution
University of Miami
521 Anastasia
Coral Gables, Florida 33146

Bernard M. Loomer, Ph.D., D.D.
Emeritus Professor
Graduate Theological Union
Berkeley, California, and
Visiting Professor, Claremont Graduate School
510 West 6th Street
Claremont, California 91711

Peter Marler, Ph.D.
The Rockefeller University
Field Research Center
Tyrrel Road
Millbrook, New York 12545

Preface

On October 4 and 5, 1977, Gustavus Adolphus College hosted the XIIIth Annual Nobel Conference in honor of the memory of Alfred Nobel. The topic for the conference was "The Nature of Life" and was a query addressing the basic tenets of the organization of living matter. From the early stages of planning through the final presentations, there was a clear demarcation of two essential questions. The first was scientific and encompassed the physical and chemical composition of the parts of life. The second was more philosophical and humanistic and dealt with the characteristics of life.

The conference focused on microscopic and macroscopic organization, including thermodynamic origins and the evolution of order from chaos. It was fitting to begin with Dr. René Dubos, the eminent microbiologist and humanist, who spoke of the "mutualistic symbiosis" of life. Impressed with the biological power of nature in its regenerative capacity, Dr. Dubos drew upon a view of life as a depiction of historical consequence. The composition of life at any particular point in time is the result of a biological history and the history of the environment in which life evolved. Life has transformed the environment and has in turn been transformed by it. This creative association and interdependence of life and the environment have allowed life to transcend what we know of molecular structures and chemistry. It is characteristic of life that it has the ability to make choices, whereas inanimate matter is governed by chance. The choice comes in some measure from the chemical mechanism involved in the origin of life, yet allows for an unpredictable evolution through creation.

Dr. Sidney Fox would challenge the unpredictability of life, at least in its origin. Dr. Fox presented a detailed analysis of the plausible events that may have led to the origin of protolife, a term used to describe the molecular organizations that occur through self-ordering principles. Careful laboratory analysis of the synthesis of proteinoid elements under prebiotic atmospheric conditions has led to the conclusion that basic chemical elements in the form of amino acids can self-order into functional protein-like compounds and can form cellular structures reminiscent of the modern cell. In this view of the origin of life, the thermodynamic characteristics of the elements of the universe have led to the very nature of the physical existence of living matter.

This physical existence is part of the unity that is observed in life, and led to the theme of Dr. Bernard Loomer's address. Dr. Loomer began with a discussion of the very method of inquiry into the nature of life. His doctrine is that evidence of truth is hard won and that all evidence is a function of perception. Echoing the creative associations of Dubos, Loomer concluded that there are no independent things in our world.

Rather, all things exist in the environment, which may be termed their "field of energy." The primacy of relations within these energy fields are the causal influences from which our existence as a self begins. The perception that is viewed is that we do not exist as individuals that relate to other individuals, but rather that we are individuals only through the relationships that are expressed. The order that the interrelationships have taken in the evolution of life is what constitutes the "web of life." The fact of the web leads to larger social structures and gives the interpretation that we as individuals are all members of one another.

Dr. Peter Marler posed the question of communication. One of the defining characteristics of living systems is their ability to communicate and thus to affect their surroundings. Dr. Marler referred to the orderly process of perception, an order imposed by strict application of physical laws. As an ethologist, Dr. Marler adds the function of interpretation to the definition of perception, and measures interpretation through behavior. Ethologists are warned that genetic bases for species differences have been overestimated in some cases of innate behavior but have been underestimated in the evolution of the human species. It is the contention of Dr. Marler that there are innate patterns of communication, as well as development of learned communication. Each has its role in the definition of life and in the survival of the individual and the species. Each also has a basis in orderly perceptions or experiences and as such fits the criteria expressed by Irwin Schroedinger for the process of thought.

Dr. Max Delbrück discussed the work of Schroedinger and included an analysis of the book *What Is Life? The Physical Aspects of the Living Cell* published in 1945. This small volume was the initial key point for the choice of topic in the XIIIth Nobel Conference, and is a philosophical inquiry into the question of whether life can be defined in purely physical and chemical terms. Dr. Delbrück forces us to ask a similar question about the nature of the mind. The approach of his presentation "Mind from Matter??" was evolutionary and suggests that mental activity is the result of biological evolution but is not bound to that level. In its formal operations, the mind can transcend the purely experiential and obtain insights that cannot be explained, but which are far from mystical. They are no more mystical than some of the very laws of physics and chemistry that are espoused currently.

Each participant in his turn focused on the perception of the nature of life that was of immediate concern to his field of endeavor. From the mixture of thoughts on the subject emerged a common thread that life is unique, orderly, and unified. There was little doubt that chemistry and physics could ultimately have a great deal to say about the composition of living matter. Whether they could complete the definition was a matter of differing opinion and depended on the concept of mind and the ability to perceive and interpret experience. Free will, in the context of biological interdependence, was left as an unknown.

The manuscripts of the participants have been faithfully adhered to in this publication, and grateful acknowledgment is given to the efforts of the participants. Their work has become a part of the tradition fostered by the Nobel Conferences, as hosted by Gustavus Adolphus College.

THE
NATURE OF LIFE

chapter 1
Biological Memory, Creative Associations, and the Living Earth

René Dubos

The Rockefeller University
New York, New York

LIFE, MEMORY, AND CREATIVENESS

Life is said to have begun on Earth when some molecules in the primeval oceans happened to replicate themselves and to continue growing. Replication and growth, however, are not peculiar to Life and are indeed phenomena of common occurrence in the inanimate world. For example, crystals replicate themselves monotonously; snowballs grow into avalanches with enormous destructive power. In contrast, Life is endlessly innovative and can be immensely constructive.

If the first truly living form had not been endowed with potentialities transcending replication and growth, it would simply have grown larger and larger, becoming a gigantic undifferentiated mass that would not fit our concept of Life. What happened fortunately is that Life continuously underwent modifications and evolved into innumerable forms increasingly different from one another. Furthermore, while thus evolving, it simultaneously transformed the terrestrial environments from which it derived its existence and progressively modified its own characteristics to fit the new conditions it had itself created. As a result of this interplay, the various forms of Life and their environments constitute integrated systems in which each component is an expression of the

1

other. The nature of Life cannot therefore be apprehended apart from the environments to which it is adapted.

Life is innovative and constructive because it incorporates its past experiences and the memory of its environments into any manifestation of its existence, thus generating diversity and preparing the ground for an unprecedented future. Some of the myths of ancient Greece seem to symbolize how memory created diversity out of the material world.

The union of Gaea (who personified the Earth) with Uranus (who personified the Sky) produced Mnemosyne who was the goddess of memory. Because the Earth had to be fertilized by Uranus to become productive, Mnemosyne can also be regarded as a symbol of Life. Mnemosyne in turn became by Zeus the mother of the nine Muses, who are the moving spirits of creativeness.

Restated in prosaic words, the Greek myth implies that the creation of novelty always involves the incorporation of past events into present conditions and therefore depends on memory. Like Mnemosyne, Life has the ability to remember the situations and events it experiences, and this attribute is the basis of its creativeness. By incorporating past events into present forms during the process of evolution, Life creates novelty in the living things themselves and also in their environments.

Because the emergence of biological novelty during evolution is now fairly well understood, the emphasis of this chapter is on the creative influence of Life on the evolution of the Earth.

FITNESS OF THE ENVIRONMENT

The fitness of an organism to its environment is an essential condition of its biological health and success. This fitness is determined chiefly by the mechanisms that enable living things to undergo changes that make them better adapted to their environments. Fitness results from the genetic changes that occur during the evolution of the species and from the physiological and anatomical changes that occur during the existence of each individual organism. In higher species, and especially in the human species, these biological processes of adaptive changes are supplemented by social adaptive changes.

Early in the present century, L. J. Henderson pointed out that the achievement of fitness depends not only on the adaptive potentialities of living things but also on certain fundamental attributes of the terrestrial environment. In his famous essay *The Fitness of the Environment*, he demonstrated that Life as we know it can exist only because the physicochemical conditions on Earth happen to be just right for its existence. In his words,

> Darwinian fitness is compounded of a mutual relationship between the organism and the environment. Of this, fitness of environment is quite as essential a component as the fitness which arises in the process of organic evolution (Henderson, 1913).

It is certain, indeed, that most forms of terrestrial life would soon be annihilated if the temperature, the salinity, the acidity, the relative proportions of gases or minerals, or any one of a number of other variables on the surface of the Earth or in its atmosphere were to sway far from their present values for any length of time. The fitness of the environment is indeed an essential condition of Life on Earth.

Henderson's concept of "fitness of the environment" is based on knowledge of the conditions required for the maintenance of Life on Earth *today*. However, it is not valid for Life in general, not even for the forms of Life that existed on Earth in the distant past. The very first living forms—from which all present forms are derived—emerged at a time when the physicochemical conditions prevailing on the surface of the Earth and in its atmosphere were vastly different from what they are now. The fitness of the environment has thus implied profound changes in the environment during the history of Life on Earth.

Furthermore, most cosmologists and biologists believe that the very activities of living things have been responsible for many of the environmental changes that have occurred on the surface of the Earth and in its atmosphere. Life itself has created in the distant past, and it continues today to create the environmental conditions suitable for its survival, its evolution, and its extension.

Biological systems are never passive in the face of threats to their existence. At all evolutionary levels, they have acted on their environments to make them compatible with their survival and further development. Admittedly, countless organisms and species have failed to meet this test and therefore have been destroyed, but they have always been rapidly replaced by other living things. All individual forms of Life are extremely delicate, but Life itself persists. This resilience is not attributable only to the readiness with which biological systems undergo adaptive changes but also, and perhaps even to a greater extent, to the fact that, in a thousand ways, biological systems can adapt the environments to their own needs. It has even been suggested that Life as a whole—the land surface, the oceans, the atmosphere—constitutes a giant system that is able to control temperature, the composition of the air and the seas, certain crucial characteristics of the soil, and other physico-chemical factors of the Earth, so as to ensure optimal conditions for the survival of the biosphere. According to Lovelock, the system seems to exhibit the behavior of a single organism, a system which probably evolved to create fitness of the environment for the forms of Life that prevail on Earth today (Lovelock and Epton, 1975).

THE LIVING EARTH

The study of fossils shows that all of today's major groups of animals and plants were already represented by recognizable ancestors some 300 million years ago. Even more impressive is the fact that microscopic structures closely related to the present blue-green algae can be found in geological formations some three billion years old. Because these organisms have a complex cellular structure, it can be surmised that they have been preceded by simpler forms. However, nothing is known of these earlier forms, and thus, at present, only speculation concerning the origin of Life is possible. The most widely held hypothesis, outlined below, does not even suggest how replication molecules became organized in the cellular structures that are the fundamental units of all present living things.

Before the advent of Life the terrestrial atmosphere contained ammonia, methane, and traces of hydrogen, but no oxygen. A shallow ocean covered most of the planet's surface except for occasional islands of volcanic rock. Electrical discharges of thunderstorms caused the methane, ammonia, hydrogen, and water to react chemically and thus brought about the production of new and more complex organic substances. From the point of view of Life's origin, nucleotides and amino acids were probably most important. Rain caused the draining of these new products into the oceans, where they progressively accumulated and formed a rich organic "soup" in which they interreacted, perhaps more readily in contact with the volcanic rocks, thus generating chemical compounds of every conceivable composition, shape, and size. After countless such accidental encounters, a molecule was eventually formed that had the magical ability to produce copies of itself and therefore gave a start to the mainstream of life.

While Dr. Sidney Fox (Chapter 2, this volume) has demonstrated a self-ordering principle that may be involved in this process, another widely accepted scenario is given by a quote from the eminent biologist S. E. Luria (1973):

> In order for life to evolve, a form had to come into being that could direct its own replication. In the living world of today that role is restricted to nucleic acids alone. Most geneticists believe that the successful starters in the early history of life were nucleic acid molecules that could help similar molecules to assemble by a process of autocatalysis. Early amino acid polymers, possibly made with the help of clay, might then have been replaced by true proteins—that is, by "meaningful" amino acid chains made by translating a nucleic acid template. But the truly creative advance—which may in fact have happened only once—occurred when a nucleic acid molecule "learned" to direct the assembly of a protein that in turn helped the copying of the nucleic acid itself—in other terms, a nucleic acid served as a template for the assembly of an enzyme that could then help make more nucleic acid. With this development, the first powerful biological feedback mechanism had come into being. Life was on its way.

At some later time, another event that also borders on the miraculous caused the replicating material to assemble itself into cells, thus providing Life with the efficiency of an organized chemical factory. A cell, concentrating within itself all the molecules needed for its growth, can reproduce itself in a shorter time than free-floating strands of DNA with their associated amino acid chains. In fact, all known forms of life now have a cellular organization.

It must be mentioned that the few laboratory observations that can be quoted in favor of the hypothesis have only a very tenuous relationship to the emergence of Life.

When the components of the initial prebiotic atmosphere are exposed in vitro to the proper radiations, nucleotides and amino acids are indeed produced, thus providing the types of organic materials on which life now depends. Life, however, has never been observed to emerge in such tests.

When amino acids are heated and then immersed in water, they may swell into hollow spheres that have some gross morphological similarity to cells and that, on occasion, may even divide in two. Such phenomena probably occurred repeatedly on the young Earth, in the lava on the flanks of erupting volcanoes, but they do not produce any structures even approaching the complexity or organization of a cell. Dr. Fox has dealt at length with this in his presentation (Chapter 2) and suggests that these spheres are the first forms of protolife.

Niels Bohr has indeed claimed that Life's organization cannot be understood in such simple physicochemical terms because of a sort of Hegelian complementarity between being alive and being taken apart. He therefore has suggested that Life might be "an elementary fact, just as in atomic physics—a quantum of action has to be taken as a basic fact that cannot be derived from ordinary mechanical physics" (quoted in Frisch, 1967). But this view has found more favor among physicists than among biologists.

Regardless of assumptions concerning the "nature of Life" and whatever may have been the chemical and morphological organization of its early forms, biological growth naturally depended at first upon the store of chemicals produced by prebiotic forces. As the primitive forms spread, they used the accumulated prebiotic chemicals and thus changed the composition of the environment. Exhaustion of nutrients eventually meant impending starvation or at least slowing down of growth—the first threat of the Malthusian catastrophe. Under these conditions, the only forms of life that could prosper were those capable of synthesizing anew the organic substances they needed for their growth by using such raw materials as carbon dioxide, ammonia, or nitrogen. Thus, Life was shaped at first by the primeval environment, but soon began to modify the conditions under which it had emerged, and then itself had to change by evolving as conditions changed.

Photosynthesis was probably the most crucial event in chemical evolution. By using solar energy, cells were capable of creating within themselves a mixture of organic substances that took the place of the ocean "soup" of the prebiotic era. Whereas at first nutrients were produced chemically in the external environment by energy originating from the sun, photosynthesis made it possible for the cell to produce them biologically within its own environment. By linking itself directly to solar radiation, Life added to the Earth new substances of its own making.

The oxygen released by the photosynthetic process progressively changed the atmosphere of the Earth, such that it eventually contained 20% oxygen. This gas screened the surface of the Earth against excessive ultraviolet radiation and made it possible for many living things to escape from their ocean shelters and to invade dry land, thereby extending the biological settlement of the Earth's surface.

Oxygen also contributed to the shaping of Life by providing those cells capable of using it with a powerful source of energy. By using oxygen animal cells achieved great mobility and became capable of actively seeking food, wherever it was, and of retreating from danger whenever necessary. Furthermore, the extra energy yielded by oxygen stimulated a new line of evolution that eventually led to thinking processes. The cells of the human brain consume roughly 10 times more energy than do ordinary body cells, and it is doubtful that they could have ever developed without oxygen. Through a long evolutionary process, Life thus did develop thinking as a result of the fact that it had learned to link the Earth and the Sun through the process of photosynthesis.

A few figures suffice to illustrate the magnitude of the contribution made by photosynthesis to the Earth's economy. In their aggregate, the various forms of green plants that exist on Earth now fix approximately 840 trillion kilowatt hours of energy per year, which accumulate in the form of biomass. Of this grand total, about two-thirds are fixed by the land vegetation, especially the forests, and one-third by the vegetation growing in water, especially in marine estuaries, in various wetlands, and in plankton areas. Surprising as it may seem, the energy accumulated per year in the biomass is approximately 12 times greater than the total annual amount of energy used by all of humankind, including its most extravagant technologies. Ultimately all forms of life and the very characteristics of the atmosphere, of the water, and of the land from which they derive their nourishment are now dependent on the energy produced by photosynthesis.

It can be said without exaggeration that practically all of the Earth's surface has thus been transformed by Life. Lichens grow on rocks in the coldest regions and at the highest altitudes; microorganisms can be recovered from hot springs and from the waters of polar areas. Albert Schweitzer tried to express this pervasiveness and power of living things by the phrase *Ehrfurcht vor dem Leben* (inadequately translated into English as "reverence for life"), which has semimystical overtones of fear before the overwhelming immensity of Life and of its power as it permeates the vastness of nature (cf. Brabazon, 1975).

This power can be apprehended in our times by observing the rapidity with which life reestablishes itself in areas that have been sterilized by volcanic eruption and thus have been returned to a condition somewhat similar to that of the prebiotic era.

In 1883, the Krakatoa island in the Malay peninsula was partly destroyed by a tremendous volcanic eruption that killed all its forms of life. Experts have estimated that the explosion had the violence of a million hydrogen bombs. The seismic wave it generated reached up to 135 feet above sea level, destroying seaside villages in Java, Sumatra, and neighboring islands. Ash and gases rose 50 miles into the sky, blocking out sunlight over a 150 mile radius. Vast quantities of pumice hurtled through the air, defoliating trees and clogging harbors. When the eruption ended, what remained of the island was covered by a thick layer of lava and completely lifeless—in some ways similar to the desolated surface of the Moon or of Mars.

The wind and the sea currents, however, soon brought back some animals and plants, and life once more took hold on the lava. More than 30 species of plants were recognized as early as 1886. By 1920, there were some 300 plant species and 600 animal species including birds, bats, lizards, crocodiles, pythons, and, of course, rats. Today, less than a century after the great eruption, the plant communities on Krakatoa are approaching the composition of the climax forest in the rest of the Malay Archipelago.

The most recent illustration of Nature's biological power is the rapid establishment of living things on Surtsey, the new island created by a submarine volcanic eruption on November 14, 1963 off the coast of Iceland. Within less than 10 years after its emergence, Surtsey had acquired from the neighboring islands and from Iceland itself a complex biota that makes it an almost "normal" member of the Icelandic ecosphere.

Many examples of such resurgence of life have been observed under other conditions. Even Bikini and Enivetok, pulverized and irradiated by 59 nuclear blasts between 1946 and 1958, are said to be reacquiring a seminormal biota, despite the destruction of their top soil. The establishment of living things on Krakatoa and on Surtsey after the volcanic eruptions offers some analogies with what happened three billion years ago. Then, as now, Life took a foothold on the bare lava almost free of organic matter, but the rapidity with which the phenomenon took place in our times is caused entirely by a set of conditions that Life itself has created. Our planet is no longer an ordinary celestial body; it behaves rather like a highly integrated living organism endowed with great powers of regeneration.

CREATIVE ASSOCIATIONS

Living things form integrated systems with inanimate matter on the surface of the earth. Moreover, they organize themselves into biological associations that can be so intimate and creative as to constitute entirely new systems of mutualistic symbiosis, a process that began very early and that occurs in many forms at all levels of organization, including even the subcellular level.

The biochemical unity that underlies the living world can be explained only by assuming that primitive organisms had already developed most of the metabolic pathways for the production and use of energy as well as for the synthesis and degradation of essential biostructures. In other words, most of the important biochemical innovations occurred very early in the history of Life, after which evolution proceeded not by creating biochemical novelties but by rearranging and using differently over and over again the same fundamental elements that govern the metabolic and structural aspects of Life. To quote Francois Jacob (1977),

> What distinguishes a butterfly from a lion, a hen from a fly, or a worm from a whale is much less a difference in chemical constituents than in the organization and the distribution of these constituents. The few big steps of evolution required acquisition of new information. But specialization and diversification occurred by using differently the same structural information.

It may well be that the first and most important type of biological association occurred through some form of recombinant DNA.

Another type of biological association began when certain cells that had grown together became specialized in the tasks they performed. All multicellular organisms consist of cells that are genetically identical but that complement each other by differentiating to fulfill different functions. The division of labor among cells made it possible for a colony to be more efficient to find food, to grow, and to avoid danger. As a result, these associations of specialized talents eventually became the dominant form of Life on Earth.

Probably more important for the extension of Life were the associations between organisms of different genetic constitutions. Plants, for example, owe their ability to fix solar energy to the presence in their cells of chloroplasts, which produce the chlorophyl responsible for photosynthesis. Genetically these chloroplasts are different from the genes of the plant cells that harbor them. They probably were initially microorganisms capable of independent life, but they now function symbiotically in the plant cells. Mitochondria, which carry out essential metabolic functions in an immense diversity of cells, also probably began as independent microorganisms before assuming their symbiotic life in association with other cells. Chloroplasts and mitochondria retain the biological memory of their independent existence, but their association with other organisms has been immensely creative and has been one of the main factors in the extension of Life.

Obligatory symbiotic associations between different species are extremely common in nature; indeed, they are probably the rule rather than the exception. Furthermore, even organisms that are capable of independent existence in artificial environments of human design are never found to live independently when they are in their natural environments. In nature all forms of life function in association with several genetically unrelated species. The characteristics of these associations are conditioned, of course, by the evolutionary past of each particular organism and by the total environmental conditions prevailing at a given time, but, under normal circumstances, the outcome of the association is favorable for all the components of the ecosystem.

A few examples, outlined below, illustrate how the creativeness of life is determined by the manner in which the biological memories of the individual components of a given ecosystem interplay with its physicochemical components.

The word symbiosis was introduced almost a century ago by DeBary to denote "a phenomenon in which dissimilar organisms live together." The particular symbiotic system for which DeBary coined the word symbiosis was that of the lichens, the various forms of which had just been recognized to consist of a particular species of microscopic alga living in close association with a particular species of fungus. The following facts show, however, that the word symbiosis hardly does justice to the creative complexity of the relationship between the two microscopic species. Alga and fungus do not merely "live together;" their association results in anatomical structures, chemical activities, and physiological characteristics that transcend the merely additive attributes of both organisms.

The symbiotic algae and fungi can readily be isolated from their association in the lichen and can then be grown independently in the proper culture media.

Combinations of certain fungi and algae have been so successful that there are some 20,000 types of lichen distributed in most environmental habitats. Even more remarkable is the fact that each type of lichen produces its own set of complex morphological structures and unusual chemical compounds that neither of the constituents—alga or fungus—could produce alone. Furthermore, all lichens exhibit the ability to survive extreme levels of heat, cold, or dryness and to multiply in the most uninviting places. In other words, the attributes of each lichen are much more than the sum of the characteristics of its algal and fungal components.

The association between alga and fungus can be disturbed readily and, indeed, destroyed completely by changing the composition of the atmosphere or of the supporting medium. For example, lichens are readily dissociated into their algal and fungal parts by atmospheric pollution or, more interestingly, by providing them with a rich nutrition. When the two symbionts are thus separated, they can be cultured separately and have been shown to have each retained its own distinctive genetic endowment. The biological memory of each has survived unaltered, despite the long and complex association in nature.

Lichens can be "resynthesized" only by placing alga and fungus in a medium so deficient that it is unable to support the growth of either one of them separately. Thus, physiological necessity is the driving force of lichen formation and existence, a truly remarkable chef d'oeuvre of evolution by symbiosis.

Higher plants and bacteria exhibit phenomena of creative symbiosis as startling and as unexpected as those just described for lichens. In nature the roots of the various species of legume plants exhibit nodules that harbor large numbers of bacteria belonging to the *Rhizobium* species. Each type of legume has its own strain of *Rhizobium*. The exquisite affinity that exists between a particular type of legume and a particular strain of *Rhizobium* is obviously the consequence of long evolutionary association and adaptation. Yet both organisms can grow in complete independence of each other if they are provided respectively with the proper nutrition and other environmental conditions. In other words, both have retained their specific biological identity despite their constant association in nature.

This association results not only in new morphological structures, which form the root nodules, but more interestingly in the production of a hemoglobin-like substance that enables the plant-bacteria system to use atmospheric nitrogen. These creative aspects of the symbiotic relationship are extremely sensitive to environmental conditions. For example, fixation of atmospheric nitrogen does not take place if the plant is richly provided with nitrogen fertilizers. Furthermore, the bacteria may not remain localized in the root nodule but instead invade other tissues and kill the plant if the latter is grown in a soil or nutrient fluid of inadequate composition. Thus, it is only when the individual biological memories of the two different organisms associate under just the right conditions that they constitute a creative system. They then become capable of enriching the Earth not only through the production of organic matter from carbon dioxide by photosynthesis but also through the incorporation of atmospheric nitrogen into this organic matter.

Animals also depend on symbiotic associations with various species of microorganisms for their growth under normal conditions. Suffice it to mention here that many animal species, from insects to the most developed mammals including humans, fail to achieve normal development of their anatomical and physiological potentialities if they are deprived of the microbial life with which they are always associated in nature.

Animals, furthermore, become organized into complex ecosystems with the flora, the fauna, and the physicochemical forces of the particular environment in which they normally live. Such ecosystems commonly exhibit great stability, but their resilience does not necessarily mean that they do not undergo changes. A common form of resilience is continuity achieved through creative change. Animal life can exert creative effects in the ecosystems of which it is a part by accelerating the turnover of organic materials and by bringing about qualitative changes in vegetation. The two following examples illustrate the role of animal life in the creation of new ecosystems.

The animal population played an essential role in the development of the great prairies on the North American continent. Until the turn of the century, immense herds of buffalo trampled open spaces in the grass and at the same time richly manured the soil. In the words of a nineteenth century observer, the buffalo "press down the soil to a depth of 3–4 feet. . . . all the old trees have their roots bare of soil to that depth." The spaces opened in the grass by the buffalo were used by smaller animals, such as prairie dogs, which not only supported predators such as the black-footed ferret but also turned over enormous quantities of earth by their incessant borrowing activity.

Whatever the factors involved in the evolution of the prairie vegetation, it is certain that the final result was a balanced system with luxurious, tall grasses, dozens and dozens of species of wildflowers, and a black sod more than 10 feet deep in certain places.

So many random events were involved in the emergence of the American prairie that it would probably be impossible to recreate this ecosystem today even if all its plants were known. The flora could not be reestablished in their original state without the participation of all the ancient fauna and other natural forces, including the trampling of the soil by immense herds of buffalo.

There are also European examples of ecosystems that acquired their characteristics through the effects of animals on the soil. During the middle of the fourteenth century, for example, large areas of arable land were abandoned as a result of the plague epidemic (Black Death) that decimated the population. If the plague had occurred in earlier Saxon times when the land was cultivated by individual settlers, many fields certainly would have been abandoned for lack of labor and therefore would have returned to the original forested state. The feudal system, however, created a different social situation. Because one shepherd with a flock of sheep could deal with much more land than could many plowmen and oxteams, the lords could keep their land in productive use in spite of labor shortage. The large flocks of sheep that were thus maintained during the fourteenth century destroyed much of the new tree growth and converted the land into pasture.

Traditional sheepgrazing, unlike the munching of cattle, cropped the grass to a lawn-like texture, which stimulated the growth of the grass and of many wildflowers, rock rose, wild thyme, scabious, etc., which now make the Sussex Downs smell, in the words of Rudyard Kipling,"like dawn in paradise." The multitude of insects, especially butterflies, also depends on these sheep-grazed plants. Thus, in this case again, the creation and maintenance of a highly desirable ecosystem is dependent on a multiplicity of random factors, which result in the inhibition of tree growth by the animal population.

HUMANIZING THE EARTH

It is commonly assumed that the impact of humankind on the surface of the Earth did not become significant until the size of the world population increased enormously in our times and particularly until the advent of modern technology. In fact, however, the most profound transformations of the earth by human activities began long, long ago, when the world population was extremely small and equipped only with fire and a few simple tools and weapons.

The process began during the Stone Age as soon as humans moved away from the semitropical savanna where they had acquired the biological characteristics that define our species and had settled over most of the Earth's surface, chiefly in climatic and topographical regions far different from those of their biological origin. The genetic constitution of the human species not having changed significantly since the late Stone Age, the immense majority of humans now live in natural environments to which they are not biologically adapted. As a consequence, they must transform these environments so as to create artificial ecosystems and habitats suitable for their own biological needs. They could not survive long even in the temperate zone if it were not for the humanized ecosystems they have created out of its primeval forests and swamps. Human life implies the humanization of the Earth.

The biological memory of the human species has guided the transformation of the primeval wilderness into artificial ecosystems and habitats throughout the world. For example, practically all our food still comes from sun-loving plants, as was the case for primitive humankind in its original savanna. This fact has necessitated clearing the forests for the creation of open farmland. Our sensorial needs also reflect our biological origins. In the savanna early people experienced a wide panorama of short grasses and scattered trees and bushes, a kind of landscape that generated visual tastes (and perhaps even visual needs) that are still almost universal today. The ideal landscapes of painters, park designers, and homesteaders have always reflected the vast horizons in which our species acquired its biological identity and its aesthetic criteria.

Uniformity of needs is fortunately compatible with a great diversity of solutions to satisfy them. Many different kinds of crops suitable for human nutrition can be grown by a wide diversity of agricultural systems, provided there is adequate insulation, temperature, and other environmental conditions. Similarly the need for open vistas can be satisfied by many different types of landscape design, for example, by the sweeping lawns of the great English estates, by the long straight avenues of trees in the French classical parks, by the openness of the best American parkways.

Granted that much intervention into nature has been destructive, it is also true that humans have created artificial landscapes all over the globe, which reveal potentialities of nature that would have remained unexpressed in the state of wilderness. The techniques used by traditional farmers to shape and maintain their lands in a productive state imply a harmonious interplay between human nature and external forces, resulting in adaptive fitness and aesthetic quality. Successful rural landscapes always imply good ecological management. Much of what we lovingly call nature has in fact been sculpted out of wilderness to create agricultural land. Some of the most diversified, beautiful, and productive ecosystems of the world have thus been transformed and maintained in a healthy state for lengthy periods of time by human labor and imagination.

Humans transform the wilderness primarily to satisfy their biological needs, but while so doing they create different kinds of environments that eventually influence their behavior and the structure of their societies. As Winston Churchill once remarked, we shape our environments, then our environments shape us (Eade, 1952). The humanization of the wilderness has thus resulted in a complex partnership between humankind and the earth, a symbiotic mutualism that creates new values in both components of the system.

THE ATTRIBUTES OF LIFE

As already mentioned, Life is probably much more ancient than the oldest forms that have been detected as fossils, but its earliest forms were certainly so delicate and undifferentiated that they have survived as identifiable fossils. In any case, lack of fossil evidence is not the only difficulty in dating the "origin" of life. A more fundamental difficulty is to judge when a self-reproducing system has attributes that entitle it to be labeled as *living*.

There is an epistemological problem inherent in the view that Life "originated" from a primeval chemical soup made up of various simple chemical compounds that formed nucleotides and peptides under the influence of random bombardment by cosmic radiations. The earmark of all living things known today is not their chemical composition, but the organization and integration of their constituent parts. Any system that we know to be living is made up of various components that are interrelated and interdependent. Because interdependence implies that one component cannot exist without the other, it would seem that living systems, as we know them, could hardly have come into existence stepwise, one component being formed and then another in such a sequential and orderly manner that they could fit and become integrated as a biological whole. Can a self-replicating molecule be called living? Or does Life imply an integrated biological system? Can one postulate that there is not an "origin of Life" but only a continuous trend toward organization and integration inherent in inanimate matter? Is it meaningful to assume, along with Niels Bohr, that Life is an elementary fact that cannot be derived from its physicochemical manifestations but runs through all living forms like a continuous thread?

While I cannot contribute to the solution of this epistemological problem I can at least express my view that the phrase "the nature of Life" denotes not only the composition and organization of living things at a given time but also their biological history and the history of the environments in which they have evolved. The nature of Life results from the history of the Earth, which has itself been profoundly influenced by the chemical activities of living things. In other words, the "nature of Life" refers to a complex memory system in which the history of the environment is as essential as the history of the evolving organisms.

Each living species reflects in its genetic endowment the environmental conditions that presided over its evolutionary development and also the conditions that shaped the organisms from which the species evolved. However, evolutionary change is not limited to biological species. Environmental conditions are transformed continuously not only by cosmic events but even more by Life itself. To mention but a few of these transformations: the chemical components of the prebiotic soup were converted into more complex organic matter by the primitive forms of Life; rocks were covered by organic matter and slowly altered by organic action; photosynthesis consumes carbon dioxide and releases oxygen into the atmosphere; atmospheric nitrogen is being incorporated into organic matter.

Thus, Life continuously and simultaneously creates both its *milieu interieur*, through adaptive changes in its genetic and phenotypical characters, and its external environment, through its action on inanimate matter. Lucretius' maxim that nothing arises save by the death of something else means more than the transfer of organic substance and attributes from one organism to another. It implies that each organism, while living and after death, contributes to the evolution of the Earth and thus to the creation of new environments from which new living forms continue to emerge.

Two recent imaginings in the field of space biology illustrate how any vision of Life, whether extraterrestrial or on Earth, must take into account the evolution of both the living forms of the environment through which Life has manifested itself ever since its beginnings.

At a Conference on Communication with Extraterrestrial Intelligence, it was suggested (in jest, of course) that the DNA nucleotide sequence of a cat be transmitted via radio signals. The assumption was that the DNA chemical formula would be meaningful to an advanced alien civilization somewhere in space, for whom receiving the formula of the cat DNA would be equivalent to receiving a picture of the cat itself. The truth is, however, that no one could possibly construct the picture of a cat *only* from knowledge of its DNA nucleotide sequence. The advanced alien civilization could visualize the cat phenotype only if it knew the precise conditions under which the cat genotype differentiates and develops; in other words, it would have to know the terrestrial environment and expecially the internal environment of the mother cat in which the cat genome is converted into the cat phenome.

The message of the story is that it is not possible to think of biological development, or to imagine the organism resulting from it, without simultaneously taking into account the genetic endowment and the total environment. Both parts of the system are indeed so interwoven that the history of both must be considered in any attempt to convey the nature and the manifestations of Life.

The structural unity of the Life-Environment system appears even more sharply in the attempts made to formulate projects for the human colonization of Mars, as described in a report, "On the Habitability of Mars: An Approach to Planetary Ecosystems," issued by the National Aeronautics and Space Administration (1976). The objective of the program outlined in this report is not to sequester the human colonists in domed cities on Mars but rather to make the planet itself fit for human life by transforming it into an ecosystem having many characteristics of the Earth.

In brief, the goal is to alter the Martian environment in such a manner that it provides the oxygen, water, moderate temperature, protection from ultraviolet radiation, and other fundamental physicochemical characteristics required for present day earthly life. This ambitious and imaginative project of planetary engineering would involve establishing on Mars populations of hardy blue-green algae, such as those found in the dry, cold valleys of Antartica, in the hope that their photosynthetic activities would provide oxygen for the Martian atmosphere. According to the NASA report, "Photosynthesis is expected to have the capability of generating oxygen in the amounts necessary to make Mars inhabitable," provided there is sufficient water available and that the microorganisms can withstand the fierce ultraviolet radiation.

Computer models indicate that, with such procedures, about 100,000 years would be needed for "unlocking the potential of Mars for human habitation." The process might possibly be accelerated by several methods including: various physical techniques for raising the Martian temperature; the use of genetic engineering techniques to create supermicrobes capable of thriving in the existing Martian environment and of displaying in it a vastly increased rate of photosynthesis. Again, in the words of the NASA report, "The entire gene pool of the Earth might be available for the construction of an ideally adapted oxygen-producing photosynthetic Martian organism." The report emphasizes, furthermore, that it would not be sufficient to introduce a single species capable of carrying out the oxygen-generating task. The maintenance of Life would require, as on Earth, the constant recycling of the various elements needed in self-sustaining biological systems. In other words, Mars could be made habitable for humans only through the creation of a complex planetary ecosystem similar to that which has developed progressively on Earth during the evolution of Life and of its environments.

Thus, the planetary engineering envisaged for the human colonization of Mars amounts to recreating on that planet the succession and integration of the biological and environmental scenarios that have generated the present "nature of Life" on Earth progressively over the past three billion years. Whether this is possible is questionable because the succession of events of Earth has occurred according to a sequence of opportunities that is probably unique and that in any case is not likely to be repeated elsewhere in the solar system. However, the problems posed by making Mars habitable for humans have had the merit of highlighting some of the characteristics of Life on Earth.

Without exception, all known living things depend on other living things and on the other components of their ecosystem for survival and development. Furthermore, the higher the organism on the evolutionary scale, the more exacting is its dependence on a complex ecosystem. One of the major trends that determines the "nature of Life" as it is known today has thus been the emergence of more and more complex ecosystems, exhibiting a high level of integration.

However, along with this trend toward interdependence and integration, there has been a simultaneous and opposite trend toward freedom and independence that also intensifies as one ascends the evolutionary scale. Evolutionary and individual development seems to be associated with the gradual insertion into Life of more and more freedom from the constraints of matter and more and more control over the external world, a trend that has reached its present peak in the human propensity to create artificial ecosystems according to imaginary visions of the future.

Even the most cursory observation of nature reveals that, while all living things are admittedly conditioned by environmental forces, reciprocally they impinge on the ecosystem in which they live and modify it for the sake of their further development. The nature of Life encompasses a wide spectrum of such relationships with Nature, from the blind, deterministic chemical reactions by which organisms transmit their genetic characteristics to their descendants and react with their environment, to the manifestations of consciousness and free will by which humans and perhaps other highly evolved organisms create within themselves a conceptual world almost independent of the external world.

There is as yet no way to link these two extreme and seemingly incompatible manifestations of Life. Yet both are real and both have been highly influential in giving to our planet its most appealing characteristics. Without biological processes, the surface of the Earth would be as desolate as the surface of the Moon or of Mars. Without human choices, the living Earth would not generate the stately order of classical gardens or even the bucolic charm of old farming settlements.

Ortega y Gasset once wrote (1941) that "Man has no nature; what he has is history." A similar statement could apply to Life. We are in the dark concerning its origin and concerning the fundamental mechanisms that make it so obviously different from inanimate matter. But we have much knowledge of its history. We can be almost certain that all its known forms have evolved progressively from a common ancestry that can be traced back more than three billion years. We know also that living things have evolved not by changes in their gross chemical composition but by what they have done to transform their organization and the environments in which they have lived. There is no doubt, therefore, that Life is historical, but the view that its "nature" is indeed its history does not settle the question of how this history was made—whether it was completely deterministic or was influenced by forces or choices that are not yet understood or even identified.

Many biologists believe with S. E. Luria (1973) that Life is entirely "determined...by the events of the past...and that the present is not the gateway to a hopeful future but the chancy outcome of past escapes." In his words, "Genes are but molecular structures, cells are but chemical factories and organisms are but the pawns of the blind master: evolution. Man is alone to know the joys and the torments of conscious will."

Although Luria may be right in stating that only humankind possesses conscious will, it seems to me that other forms of Life exhibit characteristics that transcend what is known of molecular structures and chemical factories. In fact, I am inclined to believe that the ability to make choices among alternatives is a fundamental attribute of Life. This ability becomes more obvious and more effective the higher the organism is on the evolutionary scale, but it may exist in all forms of Life even if it is not apparent to human eyes and intelligence.

Whereas inanimate matter is entirely ruled by chance and necessity, Life copes with randomness in one way or another. Organisms deal with change behaviorally by mechanisms that *appear* purely deterministic in the case of the stimulus-response processes of primitive organisms but that give evidence of choice in many animals. Even natural selection may not be as completely deterministic as commonly stated because animals make choices concerning the places in which they establish themselves. As to humans, few are those who doubt their ability to insert free will not only in their own behavior but also in ecological determinism.

By coping with randomness in a thousand different ways, either deterministically, or willfully, Life has been capable of introducing on Earth a degree of order, organization, and diversity that did not exist before its advent on the bare planet. Three billion years of biological history and millenia of human history demonstrate that by functioning in mutualistic symbiosis with the Earth, Life and humankind can invent and generate futures not predictable from the inanimate order of things and thus can engage in a continuous evolutionary process of creation.

REFERENCES

Brabazon, J. 1975. Albert Schweitzer, A Biography. G. P. Putnam, New York.

Eade, C. (ed.). 1952. Onwards to Victory: War Speeches by the Right Honorable Winston Churchill. Cassell, London. pp. 316–318.

Frisch, O. R. 1967. Niehls Bohr. Sci. Am. 216:145–158.

Henderson, L. J. 1913. The Fitness of the Environment. Macmillan Co., New York.

Jacob, F. 1977. Evolution and tinkering. Science 196:1161–1166.

Lovelock, J., and Eaton, S. 1975. The quest for Gaia. New Scientist 65: 304–306.

Luria, S. E. 1973. Life: The Unfinished Experiment. Charles Scribner's Sons, New York.

National Aeronautics and Space Administration. 1976. On the habitability of Mars: An approach to planetary ecosystems. M. M. Averner and R. D. MacElroy (eds.), NASA SP-414. National Technical Information Service, Springfield, Va.

Ortega y Gasset, J. 1941. History as a System. Norton, New York.

chapter 2
The Origin and Nature of Protolife

Sidney W. Fox

Institute for Molecular
and Cellular Evolution
University of Miami
Coral Gables, Florida

Not so long ago, the nature and origin of protolife, i.e., original life,[1] was widely considered to be an imponderable question for

This chapter is Contribution No. 318 of the Institute for Molecular and Cellular Evolution. A principal source of financial support for the research reported here has been Grant NGR 10-007-008 from the National Aeronautics and Space Administration. In addition, grants from the Rockefeller Foundation and the General Foods Corporation were especially helpful in the early stages of research. Others who aided significantly were the National Science Foundation, the National Institutes of Health, Eli Lilly and Co., the Upjohn Company, Mr. David Rose, and the National Foundation for Cancer Research.

[1]Life, according to a now widely held view, arose in stages. We not only have contemporary life, there has also been primordial life and stages of postprimordial life (Fox, 1974b) and procaryotic and eucaryotic life (Margulis, 1971). Also, the origin of life is now traceable to the dust of stars.

Despite the difficulty of defining life, and despite the absence of any human to judge it, the truly dramatic event was the separation of a microscopic unit partly independent of, and partly dependent upon, the environment (Fox, 1960).

In the past, premature definitions of life have tended to confound and inhibit the design of experiments. I believe an eventual definition will be constructed around the evolution of information. Somewhat similar to Gatlin's (1972) definition, I would define life as any organization of cells of which one or more is individually capable of assembling and converting energy and matter into an offspring that will grow to resemble the parent.

The early evolving cell possessed the information that permitted it to make the same kind of material from which its ultimate ancestor was assembled. Basically, these capabilities are functions of the properties of the appropriate kind of matter, proteinoid in the first instance, and protein in later generations.

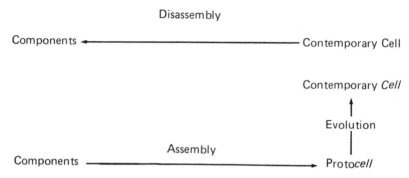

Figure 1. Current reductionistic research employs mainly disassembly. Construction-istic research is in the direction of assembly and evolution. The reductionistic research has provided clues and a standard of comparison for the constructionistic research.

science.[2] Inasmuch as the appropriate data from prebiotic matrices were not at hand, defeatism was indeed justified. The data employed in those conceptual attempts that were made were derived mainly from disassembly of contemporary organisms; this source of data also describes some that continue to be made. Such data do not explain the assembly of a protoorganism (Figure 1). In our premises and experience (Fox and Dose, 1972), what has been needed instead was the assembly in the laboratory of a de novo organism from its precellular precursor, in the forward direction of evolution itself (Fox, 1975b). Another requirement has been that each of the stages in assembly must have conformed to conditions that were geo-logically relevant at the time represented.

Research has been conducted toward the goal stated in an evolutionarily forward mode of assembly (Figure 1). The construc-tionistic hypothesis has led to experiments, which have in turn provided data from which a theory of the sequence of events has been predicated. This research has also yielded a physical product that meets most, or all, of the requirements set for a protoorganism by a number of authors (Pirie, 1954; Lehninger, 1975). The evidence for the simulated protoorganism[3] and its nature has been reviewed

---

[2]The concept that life arose once is close to the judgment that its origin is im-ponderable. This general view is still somewhat widely held (cf. Iben, 1973; Aw, 1976).

[3]In this chapter, words such as membrane, reproduction, budding, etc., are used for broader connotations than that of the contemporary cell. Some wordings in this presentation essentially signify those of candidate phenomena serving con-ceptually as evolutionary precursors for the contemporary cell.

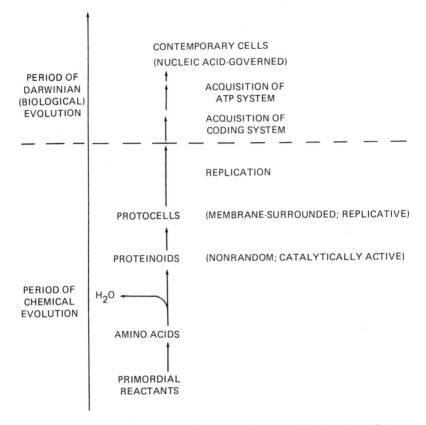

CONTEMPORARY CELLS
(NUCLEIC ACID-GOVERNED)

PERIOD OF
DARWINIAN
(BIOLOGICAL)
EVOLUTION

ACQUISITION OF
ATP SYSTEM

ACQUISITION OF
CODING SYSTEM

REPLICATION

PROTOCELLS   (MEMBRANE-SURROUNDED; REPLICATIVE)

PROTEINOIDS   (NONRANDOM; CATALYTICALLY ACTIVE)

PERIOD OF
CHEMICAL
EVOLUTION

H_2O

AMINO ACIDS

PRIMORDIAL
REACTANTS

Figure 2. Flowsheet of molecular evolution through proteinoids. Experimental simulation is fairly complete up to the horizontal line; concepts above that have no discontinuities, but the experimental demonstration is incomplete.

many times (Ambrose and Easty, 1970; Knight, 1974; Florkin, 1975; Lehninger, 1975, p. 1031; Fox and Dose, 1977). Some of the outline of the unified steps has been common in textbooks for over a decade (American Institute of Biological Sciences, 1963). The sequence of events is presented in Figure 2. This form of the chronology uses the elegant wording of Lehninger (1975, p. 1049).[4]

The flowsheet includes successive astronomical and geological matrices at its beginning, as Oparin (1924) proposed more than 50

---

[4]I would place ATP acquisition before acquisition of coding systems in Lehninger's flowsheet because of a probable need for prior ATP in synthesis of both nucleic acids and proteins.

years ago. Recent evidence from the Moon, from meteorites, and from bacteria-free terrestrial lava suggests the inevitable origin of sets of amino acids, or their precursors hydrolyzable to amino acids, in several cosmic contexts (Fox, 1973a). Laboratory experiments of Miller (1955) had already been interpreted as indicative of that beginning; those experiments, however, are regarded by some as not geologically relevant (Florkin, 1975; cf. Abelson, 1957).[5] Results of synthesis in flasks have been shown to be especially misleading quantitatively for the surface of the Moon, where there are no flasks to contain gases undergoing reaction (Fox, 1973a). Miller's experiment nevertheless qualitatively yielded a set of four amino acids found among the not more than 10 amino acids isolated from each of the terrestrial and extraterrestrial samples mentioned above (Fox, 1973a). Our assurance of the availability of amino acids derives primarily from the samples obtained from extraterrestrial and terrestrial sources, and secondarily from laboratory experiments (see footnote 5).

As Oparin and others have emphasized, the questions of the origin of life focus on questions of macromolecules and their organization.

Numerous laboratory experiments demonstrate the ease with which sets of amino acids copolymerize to yield protein-like polymers, some in the presence of minerals (Fox and Dose, 1972; Rohlfing and McAlhaney, 1976). Studies in a number of laboratories have concerned conditions of polymerization, demonstrations of catalytic activity, and self-ordering in the polyamino acids formed.

---

[5]One of the purposes of the space program is to test concepts of origins, derived from laboratory experiments. Laboratory experiments on the origins of amino acids had demonstrated significant yields of those and other carbon compounds (Miller, 1955). A large-scale analytical preparation for substantial amounts of many organic compounds was constructed (Kvenvolden, 1972; Ponnamperuma, 1972). What was found, although significant (Fox, 1973a), was very minor in amount and extremely limited in type. The fallacy was thus shown to be that of extrapolating reactions in hydrogen-rich atmospheres in flasks (Florkin, 1975, p. 247) to the open surface of the Moon. The Apollo program has thus served its purpose of disciplining concepts and experimental design.

When the small amounts of organic matter on the Moon were identified as amino acid precursors and were studied subsequently (Fox, Harada, and Hare, 1976) new cosmochemical relationships were revealed (Fox, 1973a). A degree of cosmochemical unity in the cosmos (Fox and Dose, 1977), preceding the well known "biochemical unity" was first evident (Fox, 1973a).

Of related interest are the finding of formaldehyde in a meteorite (Breger et al., 1972) and its production in the laboratory in an early prebiotic simulation (Groth and Suess, 1938).

This last effect in the overall flow of events (Figure 2) is fundamental to all of the subsequent phenomena.

Knowledge of the ease with which the thermal copolyamino acids aggregate to cell-like structures, and of the properties of the simulated protoorganisms thereby produced, is also the result of efforts in a number of research centers. Of the various functions in the microstructures, four kinds of process serving in simulated proto-reproduction have been identified: budding, binary fission, sporulation, and parturition (Fox, 1973c). In each of these processes, growth occurs by accretion of preformed proteinoid; internal cellular synthesis of polyamino acid, such as would more exactly mimic true protein, has not been demonstrated. The simulation of pre-protein and protocells have shown, however, that many phenomena could have arisen on the Earth without any direct or indirect contribution from nucleic acids.

The processes are all simple, as befits spontaneous geological occurrences. The simplicity of conditions is such that innumerable high school (American Institute of Biological Sciences, 1963) and college students (Vegotsky, 1972; Rhodes, Flurkey, and Shipley, 1975) have repeated many of the principal experiments.

Instead of another review of the evidence in detail, this chapter is devoted primarily to 1) explaining how this flowsheet comports with knowledge of longer standing, 2) discussing what has been learned more recently about underlying mechanisms, and 3) offering some newer assessments of significance.

THE NEWER VIEW OF THE PRIMORDIAL EVENTS

The theoretical understanding of the flow of primordial events has developed in stages. Three stages in sequence, derived from three types of experimental advance, stand out from the others:

1. The self-ordering of amino acids when they are polymerized by heat. This phenomenon answers a rephrased form of the chicken-egg question of biological chemistry.
2. Self-organization of the resultant thermal copolyamino acid to cell-like structures. This is the modern version of true spontaneous generation. It begins with a nonbiological source and was an event of the microscopic world.
3. The properties of the microsystems, which include some kinds of reproduction (more strictly, parent-connected multiplication).

In the research program, discovery of the first two stages noted above followed rather rapidly the rationalized construction of experiments. The third finding was highly empirical; it has required many years of stepwise investigation. Such studies cannot be expected to exhaust the potential information, much as further characterization of the contemporary cell may always be possible.

Although the prefix of "self" in words like self-reproduction deserves some qualification (Ashby, 1960), it is largely suitable for the early stages. In accord with the rudiments of evolutionary theory, each stage of development provides the matrix for the following one.

Another unexpected finding has been that the conditions for the polymerization of amino acids and the aggregation of poly-amino acids are widespread on the Earth today (Fox and Mc-Cauley, 1968).

THE CHICKEN-EGG QUESTIONS

The primary chicken-egg question, which has been with us for many years, was originally formulated from knowledge of principal components of the contemporary cell, namely, nucleic acid and protein. In the contemporary cell, each is needed for the other. The nucleic acid stores and provides the instructions for nonrandom amino acid sequence in proteins. The synthesis of nucleic acids employs protein, in the form of polymerases and other enzymes. A dilemma has therefore existed.

While the chicken-egg question is thus formulated in the context of the contemporary cell, its answer must concern primordial events. Any answer in purely contemporary terms would of course be anachronistic. Our experiments have therefore been done in as primordial a manner as could be designed. The results have yielded two responses to the chicken-egg question.

The first response is that the instructions for the ordering of amino acids in a polyamino acid need not be lodged in nucleic acids (cf. Eigen, 1971). The experiments have shown that sharply definitive instructions are obtained from the reacting amino acids themselves; this emphasis fits with the rudimentary evolutionary requirement that new phenomena emerge from matrices. Initial work on the total project of protobiogenesis was deinhibited by unexpected results from enzyme studies first reported in 1951 (Fox, Winitz, and Pettinga, 1953). Those experiments showed that the identity of amino acid derivatives synthesized by proteases was in large degree a function of the reactant amino acids. The possibility that interaction of amino acids and proteases might yield specific peptides was postulated at that time.[6] The mechanism offered may also explain the specific products in Lipmann's (1972) enzymic synthesis of polypeptide antibiotics. Lipmann's experiments employ enzymes and amino acids, which, in overall effect, dictate the sequence of the latter in the peptides.

Because the first enzyme experiments of this sort (Fox, Winitz, and Pettinga, 1953) seemed to emphasize the contribution of a directive effect of the reacting amino acids, the question arose of whether the amino acids alone, reacting under the nonspecific agency of elevated temperature and without enzymes, would order themselves. Ordering has been found under these simple conditions. Neither nucleic acids nor enzymes were necessary. Furthermore, the degree of self-ordering observed has been unexpectedly high; it provides an answer to the chicken-egg question. While a number of laboratories have recorded, in various ways, the self-ordering effect of α-amino acids when heated to combine to form polymers (Dose and Rauchfuss, 1972; Fox and Dose, 1977; Melius, 1977), this chapter focuses on three of the more recent kinds of evidence.

Dose and Zaki (1971) heated heme with mixtures of as many as 20 amino acids. After one step of purification by dialysis, these authors obtained on discgel electrophoresis a single band of "hemo-proteinoid" (Figure 3) when they experimented at each of two pHs, 8.6 and 4.5. The heme-containing polymer had a molecular weight of about 18,000, peroxidase activity many times that of the contained heme, and, as Figure 3 demonstrates, strikingly limited heterogeneity. Limited heterogeneity would, by definition, not have been possible with unordered sequences of amino acids.

Sequences of amino acid residues per se were examined in a more recent test of the self-ordering phenomena, by Nakashima et al. (1977). Nakashima heated glutamic acid, glycine, and tyrosine (Figure 4). The product was fractionated on Sephadex and on paper. The main components of the fractionation are shown in Figure 5. Essentially only two tyrosine-containing tripeptides are found: pyroglutamylglycyltyrosine and pyroglutamyltyrosylglycine.

---

[6]Another finding in extension of this observation was the dominant effect of amino acid residues in proteolysis when new methods made it possible to study hydrolysis of a protein, lysozyme, by various proteases (Hurst and Fox, 1956).

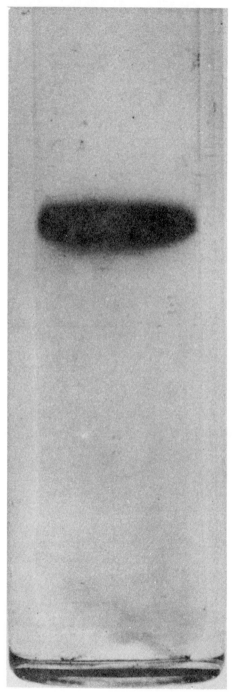

Figure 3. A single band obtained for once-dialyzed hemoproteinoid on discgel electrophoresis at pH 8.6. A similar result was obtained at pH 4.5 (Dose and Zaki, 1971).

Figure 4. Polymerization of glutamic acid, glycine, and tyrosine by heat.

The two tripeptides are present in an equimolar complex. This complex is by far the most abundant single product of the fractionation. The fact that the components of the complex are present in an equimolar ratio suggests a kind of primitive template.

The peptide product is compared in Table 1 with the tyrosine-containing tripeptides that are expected on the basis of an a priori random synthesis (Nakashima et al., 1977). The result is highly nonrandom, relative to an equal probability for each of the tripeptides that can be written. The amount of tripeptide complex isolated is calculated to be 19.2 times as great as would be expected on the assumption of unqualified randomness, or equal probability of sequences resulting from chance events.[7] The calculated probability that 19.2 is a chance result is much smaller than 1 in 10^6.

Even within the group of peptides that have the N-terminal pyroglutamic acid fixed, there are four other pyroglutamyl, tyrosine-containing tripeptides possible on an a priori theoretical basis. Of the total of six such possibilities, only the two found are present in significant quantities.

In the other recent study, Melius and Sheng (1975) condensed six amino acids (ala, glu, gly, leu, phe, pro) by heating. Upon fractionating by paper chromatography, three peptides were found. The molecular weights, the N-termini, and the C-termini of these three peptides are presented in Table 2. The most striking result is that all three peptides have only a single type of N-terminal amino acid residue, pyroglutamyl. Another notable result is that each of the three peptides is found to have but a single C-terminal amino acid. Furthermore, the C-termini of the peptides differ in identity; they are glycine, alanine, or leucine, respectively. Of the six amino acids available, only three are found in the C-terminal positions.

---

[7]Even if the replication of naked DNA could be experimentally shown, this would not satisfy the need for numerous other catalytic activities in protometabolism. At present, this is explained by the panoply of proteinoids and their easy origin.

The pyroglutamic acid thus appears to be an $NH_2 \rightarrow COOH$ polymerization initiator. The simplest way in which we can explain only three peptides in the Melius-Sheng product, with each having but a single C-terminus, is by transfer of that selectivity from the N-terminus through the peptide chain to C-termini, in some as yet unspecified manner.

With glycine and tyrosine, a similar selectivity is seen in Nakashima's results. In both the Melius and Nakashima studies,

SUBFRACTIONS

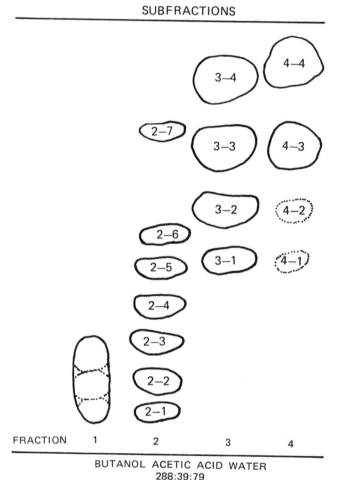

FRACTION   1   2   3   4

BUTANOL ACETIC ACID WATER
288:39:79

Figure 5. Separation of peptides from poly (glu, gly, tyr) by paper chromatography. Numbers are stoichiometric ratios of glu, gly and tyr, respectively. In Fraction 3, from the top are: dipeptide, tripeptide, tetrapeptide, pentapeptide. Fraction 3-3, the most abundant, is a complex of two tripeptides.

Table 1. Tyrosine-containing tripeptides expected from random polymerization of glutamic acid, glycine, and tyrosine and those found

A priori expectations	Found
α-Glu-α-glu-tyr	
α-Glu-γ-glu-tyr	
γ-Glu-α-glu-tyr	
γ-Glu-γ-glu-tyr	
< Glu-α-glu-tyr	
< Glu-γ-glu-tyr	
α-Glu-gly-tyr	
γ-Glu-gly-tyr	< Glu-gly-tyr
< Glu-gly-tyr	
α-Glu-tyr-glu	
γ-Glu-tyr-glu	
< Glu-tyr-glu	
α-Glu-tyr-gly	
γ-Glu-tyr-gly	
< Glu-tyr-gly	< Glu-tyr-gly
α-Glu-tyr-tyr	
γ-Glu-tyr-tyr	
< Glu-tyr-tyr	
Gly-α-glu-tyr	
Gly-γ-glu-tyr	
Gly-gly-tyr	
Gly-tyr-glu	
Gly-tyr-gly	
Gly-tyr-tyr	
Tyr-α-glu-glu	
Tyr-γ-glu-glu	
Tyr-α-glu-gly	
Tyr-γ-glu-gly	
Tyr-α-glu-tyr	
Tyr-γ-glu-tyr	
Tyr-gly-glu	
Tyr-gly-gly	
Tyr-gly-tyr	
Tyr-tyr-glu	
Tyr-tyr-gly	
Tyr-tyr-tyr	

< = pyro

the N-terminus is a single species, pyroglutamyl. It appears that pyroglutamic acid reacts more rapidly than other amino acids, including glutamic acid (Harada and Fox, 1958) and that the products are more stable with such an N-terminus.

Table 2.   Characteristics of the three polypeptides obtained by heating six amino acids and fractionating the product

	Molecular weight[a]	N-Terminus[b]	C-Terminus
Peptide-1	5,200	Glu	Gly
Peptide-2	10,000	Glu	Ala
Peptide-3	11,500	Glu	Leu

[a]By gel exclusion.
[b]No N-terminus until pyroglutamyl residue opened by heating with trifluoroacetic acid.

Our concepts are now freed of the necessity for a nucleic acid blueprint to have been translated into the first amino acid sequence on Earth. Those ordered amino acid sequences were, conceptually, the product of the self-instructing amino acids themselves (the word "self-instructing" is borrowed from Eigen (1971); it is used here in a simpler context).

The idea that the first informational macromolecule on Earth was not nucleic acid is not new, having been expressed in various ways (Oparin, 1957, p. 284; Fox, 1959; Lederberg, 1959; Ehrensvärd, 1962; Hanson, 1966; Jukes, 1966; Eigen, 1971). Thus, the first response to the chicken-egg question is the inference that nucleic acid did not arise first and that proteinoids (not protein) were the first informational macromolecules. Nor did proteinoids need to duplicate themselves; the alternative, as shown by experiment, is that the evolutionary matrix yields similar products (cf. Orgel, 1968).

A second response is that the question might best be rephrased, in the context of primordial events as we now understand them. The new emphasis is neither nucleic acid–first nor protein (or proteinoid)–first. The total array of experiments indicates that the emphasis should be *cells–first*. In any case, knowledge of the evolutionary sequence of developments in the cell is what we are interested in for the most complete overview. This emphasis has required recognition of several tenets from biology and evolution. First, recognition that the first, nonnucleated cell itself underwent evolution was necessary (Fox, 1965; Lehninger, 1975, p. 1045); the corollary is that the protocell lacked some of the qualities of the contemporary cell. This view proposes, also, that metabolism evolved synergistically with the evolution of the cell. Especially needed for conceptualization was recognition of the advantages that a cellular membrane structure confers upon evolution. Analysis of these advantages indicates that they were substantial, if not crucial (Fox, 1976b). With this definition of a protocell, in the light of the extreme ease of formation of a membrane-like structure, the answer to the chicken-egg question becomes cells–first. Nucleic acid and true protein[7] would have emerged later (Calvin, 1962; Nakashima et al., 1970), probably concurrently, and within the minimal cell (Ehrensvärd, 1962).

The earlier idea that "original protein" (proteinoid) had to be distinguished from current protein (Fox, Harada, Vegotsky, 1959) is now defined more precisely, in light of the inferred evolutionary sequence in which proteinoid yielded an intermediary protocell that synthesized true protein (Nakashima et al., 1970). The newer view is that the proteinoid served as an initial informational macro-molecule (Figure 2), having arisen as a product of favored (Fox, 1967) rather than of random,[8] polymerization (Fox, 1973b).

The history of biochemical evolution appears to include a history of modulation of reversible bidirectional reactions (of a kind that constitute organic chemistry) to unidirectional pathways of reaction sequence (of the kind that constitute biological chemistry (Lehninger, 1975, p. 373)). This emergent unidirectionalism may apply as well to recognition between amino acids and nucleotides. The recognition has been expressed by the polymeric forms of these two types of compound. The result is a consequence of molecular cooperativity. Such recognitions have been selective in both directions (Fox, Lacey, and Nakashima, 1971; Fox, 1974a). These results are consonant with theoretical considerations of "reciprocity," as analyzed in another perspective on polypeptide–RNA interactions (Carter and Kraut, 1974). Bidirectionalism in recognition could, then, have evolved to the unidirectionalism of the Central Dogma through the replacement of a kind of *reverse translation* (Fox and Dose, 1972), by the action of erstwhile new physical systems such as protoribosomes (Fox, Jungck, and Nakashima, 1974).

An alternative to explanations based on reverse translation is another that states that the forces operative in the thermal poly-

---

[8]In this chapter the statistician's definition of randomness is used: "A random process is one in which only chance factors determine the particular outcome of any single trial of the experimental operation" (Hays, 1965).

condensation of amino acids are similar to the forces that control the expression of the genetic code (Fox, 1975a). This view has molecular logic, in that individual assemblies of nucleotides correspond to individual amino acids, the whole relationship being a code. A choice between these two explanations is a choice between: 1) serial forces, i.e., forces operative at different stages of evolution with the latter expression determined by the earlier expression, and 2) sets of parallel forces, operative in the same general way but independent of each other, at different stages in an evolutionary line.

Either explanation (cf. Chernavskii and Chernavskaya, 1975; Mikelsaar, 1975) is consistent with the observed phenomenon of nonrandom amino acid sequences. Without an appreciation of the power of self-ordering of monomeric amino acids, I believe that an answer to the question of the origin of life may have been postponed indefinitely.

I also believe that, historically, the chicken-egg question has been asked at an inappropriate level. As usually phrased, it is a question of macromolecules: nucleic acid and protein. It is easy to adopt an a priori hierarchical position that many or all molecules had to precede cells, which are more complex than the molecules of which they are composed. This assumption neither honors the view that the first cells evolved (Figure 1) in concert with their components and with their biochemical and physiological functions, nor is it borne out by experiments.

Another way to deal with the hierarchical question is to ask another version of the chicken-egg question: Which came first, protein or cell (Blum, 1955)? This phrasing of the question is not as common as the nucleic acid–protein variation. A viable answer is, however, very much the same. A protein-like polymer, proteinoid, arose first. This polymer aggregated, in the presence of water, to a primordial cell. It is the primordial cell that evolved to, or was intrinsically able to, make a more contemporary type of both protein and nucleic acid.

To recapitulate, interpretation of experiments emphasizes the great simplicity of the process by which the first cells (themselves complex) could have arisen. That simplicity, taken in conjunction with the manifold and interwoven advantages of a cell-like structure for evolution (Fox 1976b), indicates that a cellular line of evolution would have had great selective advantage over any sequence that might be postulated to involve de novo cell formation at a late stage in primordial evolution. The one continuum that has been visualized from the experiments begins with an incompletely evolved cellular structure (a protocell). That protocell required as a matrix a protein-like polymer, which was intrinsically morphogenic. The proteinoid identified by experiments performed in the evolutionary direction was, according to the experimental results, eminently suitable to form a protocell; more fully evolved polymer, e.g., cellular protein, as we know it, appears by experiment not to have been quite suitable.

THE WIDER SIGNIFICANCE OF
THE SELF-ORDERING OF AMINO ACIDS

The significance of the self-ordering of amino acids extends beyond the chicken-egg question. For example, this principle demonstrates how information[9] latent in the environment (mixed amino acids) could have been the informational matrix bridging to the first organism. Informationally, the missing link between environment and protoorganism was the macromolecule ordered by its monomers.

Analysis of the influence of pyroglutamyl residues begins to provide mechanistic understanding of how copolyamino acids could have been the first informational macromolecules. However, other amino acids must also have selectively reactive features, of steric and electrostatic types (Fox and Dose, 1977). These features in amino acids are sufficiently different from those in the classical monomers of the polymer chemist, monomers such as butene and methyl methacrylate, to explain the unique vividness of the results from amino acids.

How much order was necessary in the protobiomacromolecules is an unanswered question. It may remain so because of the now recognized practical difficulty of obtaining a disordered copolyamino acid. Theoretically, some order was necessary, if only to help explain significant degrees of catalytic activity in macromolecules. In contemporary organisms, such as *Escherichia coli*, an average of many to hundreds of copies of each enzymically active molecule is deemed necessary (calculated from data in Lehninger,

---

[9]In this chapter, information is defined as the capacity of a molecule or system to interact selectively with other molecules and systems.

1975). The number of any one type in a protocell should have been even greater because of the weaker activities that the experiments indicate were available in the first polyamino acids possessing such powers. One other benefit of ordered polymers would have been that they preceded ordered protocells (cf. Lehninger, 1975, p. 1047).

An additional benefit of self-ordering relates to the fact that recognition of that possibility (Fox, 1974a) indicated that meaningful experiments in amino acid polymerization could be performed. The hope of learning by such experiments derived from the principle of the "unity of biochemistry." Because contemporary life is biochemically so much alike from one organism to another, we were led to believe that primordial life was not imponderably different. Clues from contemporary organisms could thus be used for heuristic experiments.

One more benefit of a conceptual type relates to our understanding of the number of times life began on Earth. Some infer that life arose but once on Earth, as the first event of a consequently unified biochemical world. In fact, however, self-ordering may constitute the basic reason for the unity of biochemistry from its origins. Each de novo origin, favoring certain reactions in its matrix, would accordingly have yielded products of similar composition and appearance (Fox and McCauley, 1968).

If we entertain the proposal of Copeland (1936) that life began in hot water, we can easily understand the common biota at the numerous hot springs areas. Thermophilic blue-green algae and sulfur bacteria are common to each of the many hot springs areas. The principle of self-ordering of molecules during de novo generation of organisms would account for the observed common biota.

FORMATION AND PROPERTIES OF PROTEINOIDS

The efficient thermal copolycondensation of α-amino acids is made possible by a sufficient proportion of nonneutral amino acid, the proportion of which can be small, e.g., < 25 mole percent (Fox and Waehneldt, 1968; Oshima, 1968). Carothers (see Mark and Whitby, 1940) produced an analog of protein, nylon, by heating dicarboxylic acids with diamines to yield repeated amide bonds in the same polymer molecule. He might have produced ordered co-poly-α-amino acids if he 1) had heated α-amino acids together and 2) had included a sufficient proportion of trifunctional α-amino acid.

The temperature need not be above 65° C (Rohlfing, 1976) and can undoubtedly be lower in longer periods of time (Fox, 1976a). The heating can begin on a solution of amino acids (Snyder and Fox, 1975). The warmth can both drive off the water and cause polymerization of the amino acids in the dry residue. Amino acids, being organic salts, are less volatile than water or other organic materials. Accordingly, the amino acids are concentrated as evaporation occurs. Their polymerization products are even less volatile. Inorganic salts are also generally not volatile, but the presence of salts of sodium, magnesium, etc., does not interfere (Snyder and Fox, 1975; Rohlfing and McAlhaney, 1976; Fox and Dose, 1977).

Of course, the fact that as many as 18 types of amino acid are polymerized simultaneously in the production of proteinoids does not signify that 18 amino acids were present for the protopolymerization. It does, however, illustrate the scope and versatility of the process. There are indications that interactions of several types of amino acid within the same polymer may yield unique properties not found in copolymers of two amino acid types; this suggestion is being explored.

CHEMICAL PROPERTIES OF PROTEINOIDS

The answer presented for the chicken-egg question emphasizes the degree of order in the internal structure and the limited heterogeneity of the resultant collective mass of thermal proteinoid. In addition to limited heterogeneity (Fox and Dose, 1977), the polymers have, qualitatively, some kinds of biological property found in isolated contemporary proteins. They also have the property of morphogenesis, without which a unified theory of the origin of protolife can hardly be constructed.

The enzyme-like activities of thermal proteinoids (Table 3) are, for the most part, several orders of magnitude less than those of contemporary enzymes. Aside from this quantitative difference, virtually all of the properties of enzymes have been found in thermal proteinoids (Table 4). Various proteinoids have displayed, for example, pH-activity curves, specificities for substrates, Michaelis-Menten kinetics, etc. The catalysis by proteinoids is ordinarily very much, or infinitely, greater than that of the monomers. In one kind of activity a necessary three-dimensional conformation and an active site have been demonstrated (Rohlfing and Fox, 1967).

The catalytic activities are in many cases the properties of most or all of the macromolecules of the whole polymeric product

Table 3.   Enzyme-like and related activities in thermal polyamino acids

Reaction, function, or substrate	Authors (year)
Hydrolysis	
p-Nitrophenyl acetate	Fox, Harada, and Rohlfing (1962)
	Rohlfing and Fox (1967)
	Noguchi and Saito (1962)
	Usdin, Mitz, and Killos (1967)
ATP	Fox, Wiggert, and Joseph (1965)
p-Nitrophenyl phosphate	Oshima (1968)
Decarboxylation	
Glucuronic acid	Fox and Krampitz (1964)
Pyruvic acid	Hardebeck, Krampitz, and Wulf (1968)
Oxaloacetic acid	Rohlfing (1967)
Amination	
α-Ketoglutaric acid	Krampitz, Baars-Diehl, Haas, and Nakashima (1968)
Deamination	
Glutamic acid	Krampitz, Haas, and Baars-Diehl (1968)
Oxidoreductions	
H_2O_2 (catalase reaction)	Dose and Zaki (1971)
H_2O_2 and hydrogen donors (peroxidase reaction)	Dose and Zaki (1971)
Combined enzyme-like activities in a proteinoid microparticle	Hsu and Fox (1976)
Photoactivated decarboxylation	
Glucuronic acid	Wood and Hardebeck (1972)
Glyoxylic acid	
Pyruvic acid	
Synthesis with ATP (in proteinoid microparticles)	
Internucleotide bonds formed	Jungck and Fox (1973)
Peptide bonds formed	Fox, Jungck, and Nakashima (1974)
Hormonal activity	
MSH	Fox and Wang (1968)
	Bagnara and Hadley (1970)

(Oshima, 1968), rather than being found in highly active fractions diluted by inactive macromolecules. Each substrate activity found in a thermal proteinoid is greater in some contemporary enzyme, usually by orders of magnitude, as stated. Zuckerkandl and Pauling (1965) have pointed out that the evolutionary production of enzymes as powerful as the evolved ones may well have required

Table 4.   Enzyme-like properties found in proteinoids

Properties common to contemporary enzymes and proteinoids
Molecular weight
Heteropolyamino acids
Prosthetic groups
Inactivitability
pH-Activity curves
Michaelis-Menten kinetics
Inhibition and reversal
Specificities
Relationships permitting metabolism

synthesis at a template (Fox, 1974a). The thermal experiments indicate that nucleic acid-free synthesis permitted the production of catalytically active protein-like polymers sufficiently ordered that they were capable of aggregating to a supramolecular micro-structure. The necessary numbers of molecular multiples to yield catalytically·active fractions sufficient for a protocell were attained through the self-ordering properties of the amino acids. These polymers then aggregated, in a saltatory step, to yield cell-like structures in which synthesis of powerful enzymes (and cellular reproduction) could evolve. The contemporary enzymes would have resulted from templated synthesis within cells because of the superior evolvability of a template-controlled synthesis.

RANDOMNESS[10] AND REALITY

(Interactions between Proteinoids and Polynucleotides)

A special class of chemical behavior of proteinoids is that of inter-actions with polynucleotides. Analogous, or homologous, to the amino acids' combining selectively, their polymers are also capable of binding selectively (Yuki and Fox, 1969; Fox, 1974a). This macromolecular selectivity is probably a derivative of the inter-monomeric selectivity. These relationships are believed not only to lie at the base of the answer to the chicken-egg question but also to underlie the explanation for the existence of the genetic code.

In 1968, Crick published a paper entitled "The Origin of the

---

[10] Refer to footnote 8 for definition of randomness, as used in this chapter.

Genetic Code." In that paper Crick posed two alternative explanations for the origin of the coded relationship between oligonucleotides and amino acids. One was the "frozen accident" hypothesis, which implied, in its extreme form, totally chance allocation of codons to amino acids. According to this view, correspondences between amino acids and codons were frozen into a single code and departures were selected against. Crick pointed out that this concept was not amenable to experimental testing. He suggested an alternative concept—that the correspondence between the amino acids and the (oligo)nucleotides is stereochemical, a possibility that he indicated to be amenable to experimental demonstration.

At the time acceptance of that concept was subject to the difficulty posed by the fact that interactions between amino acids with mono-, oligo-, or polynucleotides had not been demonstrated by researchers. How to overcome this difficulty was a problem upon which we had been working. Our work was biased by a belief in a stereochemical bias between nucleotides and amino acids, expressed through their derivatives or polymers (one variation of Crick's "stereochemical alternative").

The first step in this study was to determine if the molecular cooperativity theoretically inherent in polymers of amino acids might make such polymers more interactive with polynucleotides than were the monomeric amino acids. Waehneldt (Waehneldt and Fox, 1968) found that interaction could indeed be demonstrated by the formation of phase-separated microparticles. As expected, the interaction with polynucleotides was found to require a sufficiently basic proteinoid. Phase-separated particles were obtained by including at least 15–20% lysine in the proteinoid to make it sufficiently basic. The resultant nucleoproteinoid microparticles were sensitive to pH and salt concentration, much as are nucleoprotein organelles. Deoxyribonucleoproteinoid complexes tend to be fibrous and ribonucleoproteinoid complexes tend to be globular (Waehneldt and Fox, 1968), analogous to their contemporary counterparts, as in the deoxyribonucleoproteins of chromosomes and the ribonucleoproteins of ribosomes, respectively.

In the next study, proteinoids rich in arginine and others rich in lysine were allowed to interact with varied (enzymically synthesized) homopolyribonucleotides (Yuki and Fox, 1969). The results exhibited selectivity (Table 5). The pyrimidine polynucleotides interacted sufficiently with lysine-rich (arginine-free) proteinoid for particles to separate, whereas the purine polynucleotides

Table 5. Comparison of lysine-rich proteinoid with arginine-rich proteinoid in binding with polynucleotides

	Turbidity[a]	
Polyribonucleotide	Lysine-rich (arginine-free) proteinoid	Arginine-rich (lysine-free) proteinoid
Poly(C)	0.253	0.002
Poly(U)	0.050	0.058
Poly(A)	0.001	0.060
Poly(G)	0.003	0.218
Poly(I)	0.003	0.248

[a]measured at 600 nm.
Proteinoid concentration, 1.0 mg/ml; polynucleotide concentration, 0.1 μmol/ml; 0.05 M Tris buffer; pH 7.0; 25.0°C.

failed to do so. Almost conversely, the arginine-rich (lysine-free) protenoids interacted richly with the purine polynucleotides (poly(G), poly(A), or poly(I)).

The selective, i.e., stereochemical, preferences between nucleotides and amino acids are expressed for these monomers interacting in their polymeric forms. Moreover, the selective recognition is expressed in both directions: from nucleotide to amino acid, and from amino acid to nucleotide. This bidirectionalism may have evolved to unidirectional contemporary translation.

In unpublished work, Lacey (Lacey, Stephens, and Fox, 1976) has pursued the systematic study of homopolynucleotide-proteinoid interactions. His results have favored interactions of poly(C) and glycine, of poly(G) and proline, and of poly(U) and lysine, but not of poly(A) and phenylalanine. The first three are compatible with an anticodonic bias, but not the last.

In published work (Nakashima and Fox, 1972), some experiments that throw light on the relationships in another way have been recorded. Nakashima has systematically studied interactions of preformed radioactive aminoacyl adenylates (using the homocodonic amino acids glycine, lysine, phenylalanine, and proline) with simulated protoribosomes composed of each homopolyribonucleotide with one lysine-rich proteinoid. These experiments were designed on the premise that polynucleotide influences expressed during formation of a peptide bond, or of an immediate precursor of the peptides, would be most meaningful. Although the nature of the binding (or covalent reaction) of the adenylate with the particles has not been studied, regular selective patterns of incorporation can be obtained.

One set of results repeatedly obtained fits into a codonic mode. Exploration of various possibilities indicated, however, that results suggesting a matrix for codonicity, or anticodonicity, or some other relationship could be obtained. The particular set depended upon variable factors from one experiment to another, e.g., proportion of lysine in the proteinoid, ratio of polynucleotide to polyamino acid during production of the particles, etc. We have interpreted this to mean that numerous molecular matrices for genetic codes, and subsequently genetic codes, were possible in the evolution of the nucleic acid-genetic mechanism, prior to the time one was selected.

The inferences drawn to this point deserve the reemphasis that the simulation is heuristic and necessarily incomplete. Other components of the evolved genetic apparatus, such as the synthetases, probably played significant roles that have not been simulated by the experiments done so far. However, the experiments to date do lead to the inference about the selection of a single genetic code from what otherwise might have been many. This inference involves feedback through early cellular generations (Fox, 1971).

As one step in the explanation, the recognition of nucleotide identity in polynucleotides is a function of, among other specifications, the lysine content of the proteinoid (Fox, Lacey, and Nakashima, 1971). Conversely, the lysine content of protoproteins would have been determined by the composition of the protogenome. In this way, we can visualize that a maximal reflexive fixation of 1) lysine content in any one protein and 2) nucleotide composition might have been attained in only a few generations. It is difficult to imagine that such feedbacks would not have developed. The actual processes, which must have been more elaborate and complex, would have conceptually been subject to such fixation.

Many more experiments are needed in this area. However, the "stereochemical alternative," to use Crick's term, now seems to be indicated to be the real basis for the genetic code. An additional suggestion from the experiments is that the relationships were "frozen," by *feedback fixation*. Because specific proteinoids were products of stereochemical forces, and because their interactivity was also stereochemical, it can be inferred that the occurrence that was "frozen" was not from a playing-card type of randomness (see footnote 8) in the array of possibilities. Since it seems likely that the nonrandomness of proteinoid-polynucleotide interaction is a derivative of the nonrandom polymerization of amino acids, it was also not an "accident." Interactions of polyamino acids and polynucleotides are selective, analogous to the reactions of amino acids with each other. Irrespective of the interpretation, the selectivity is an experimental observation.

ASSEMBLY OF THE PROTOCELL

The self-assembly of a first cell was recognized as a possibility almost a century before the reactions of amino acids to yield self-ordered polymer were explained by experimental demonstration. Pasteur (cited in Vallery-Radot, 1922) is alleged to have said,

> There is the question of so-called spontaneous generation. Can matter organize itself? In other words, are there beings that can come into the world without parents, without ancestors? That is the question to be resolved.

Although Pasteur's experimental contributions did not test the concept of self-assembly of appropriate matter to cells, he allowed for this possibility with the question, "Can matter organize itself?" His statement of the problem stands as the most effective definition of it to this date.

The power of self-assembly seems to have been first vividly recognized about the middle of this century by Schmitt (1956), who caused collagen to aggregate into beautiful structures. Invoking Schmitt's experiments, Wald (1954) suggested that self-assembly played a role in the origin of life. Since Schmitt's demonstration the biological power of assembly of ampholytic macromolecules has become a principle of biochemistry (Fox, 1968; Lehninger, 1975, p. 1011).

While the concept of the self-organization of matter to form Pasteur's "beings," i.e., protocells (in this author's view), is not new, the organizational process for a polymer of amino acids of nonbiological origin has some special aspects. Although the product of this process is highly complex, the process is operationally simple. It occurs with high efficiency from various copolyamino acids brought into contact with water. The process is rapid, and it would have occurred on the Earth wherever a set of amino acids were present in a warm locale (Rohlfing, 1976), after which water could have come into contact with the polymer formed. The laboratory process is depicted in Figure 6.

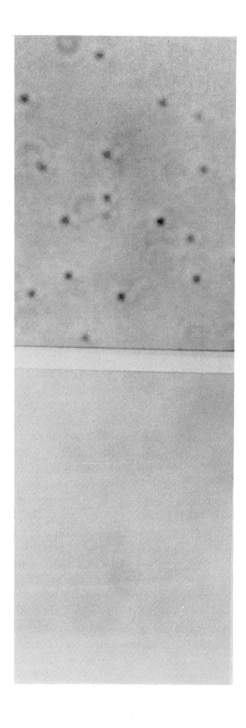

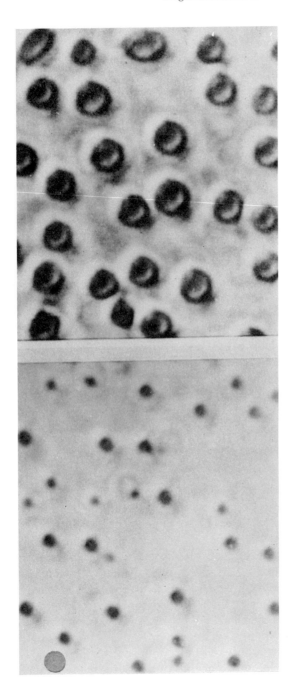

Figure 6. Assembly of proteinoid microsystems from a cooling saturated solution. The entire sequence occupied less than 2 min. No. 3 was photographed 30 sec after No. 1. ×2,500.

The occurrence of this assembly during a warm-to-cold treatment was postulated (Fox, 1957) in advance of the experiment, for the wrong reason. The assembly can occur in either cooling hot water or directly in cold water (Fox and Yuyama, 1963b). The temperatures employed in the warm-to-cold experiments, e.g., 40° → 25° C, conform to a diurnal temperature variation.

We now understand the mechanism of the organization of the simulated protocells as one based on the powerful binding ability of the polyampholytic proteinoid molecules. This binding appears as a nuisance in chemical studies of fractionation of proteinoids from mixtures with other molecules, but strong binding power may have been crucial to protobiogenesis. We believe that it is when the solubility of proteinoid in water is exceeded that the molecules in the phase-separated material bind to each other to form cell-like structures. The conditions suggest that this process of spontaneous formation of protocells would have occurred innumerable times in innumerable locales on this planet during its long history.

Simple variations of the experiment, with complex consequences, are known. Microsystems prepared from the heating of seawater salts with the usual mixtures of amino acids have much stability at high pH or high temperature (Snyder and Fox, 1975). This seawater proteinoid is composed of basic and acidic proteinoids (Figure 7), a pair which had earlier been shown to yield microsystems stable (Fox and Yuyama, 1963a) in the pH range presumed for the primitive ocean.

FUNCTIONS AND EVOLUTION
OF PROTEINOID MICROSYSTEMS

The cell-like properties of the proteinoid microsystems are numerous; their concerted association within individual units represents a principal aspect of the proteinoid theory (Figure 2). Although the ordered thermal polyamino acids were from a reasoned experiment, and the aggregation of the polymers was rationalized in advance (Fox, 1957), albeit in a theoretically incorrect manner, the properties of the microsystems were uncovered in an essentially empirical manner. Some of the properties, e.g., budding, double layers in the membrane, and patterns of association, simply called attention to themselves upon simple observation. Some other properties were sought when their presence was suggested by the composition of the units and the properties already catalogued in the research.

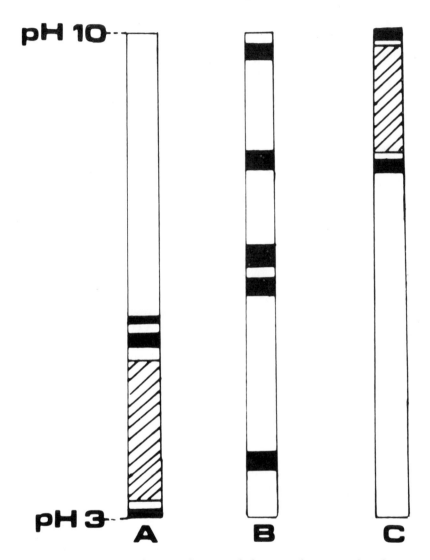

Figure 7. Gel isoelectric focusing of proteinoids from mixed amino acids and sea-water salts by heating. Both acidic and basic proteinoids are found to have formed; there are only a few of each.

If the proteinoid microsystems meaningfully simulate the protoorganism, as some authors suggest (Kenyon, 1974; Knight, 1974; Fox, 1975b; Lehninger, 1975), the description of these units is a recital of the nature of the protoorganism and comports with the title of this chapter and that of the XIIIth Nobel Conference.

However we are unable to draw rigorous conclusions because we do not knowingly have any living protoorganisms at hand for comparison.

Historically, investigators knew about contemporary cells before they conceived of protocells. Contemporary cells provided the clues to what was needed for the protocell. Subsequently, contemporary cells provided the standard against which progress in research on the forward evolution from polyamino acids could be roughly measured.

In addition to being a kind of investigation essential for understanding the emergence of life, this kind of research has the advantage that the investigator can single out one kind of component of the contemporary cell, in this case protein, can produce something quite like it under geological conditions, and then can observe which functions are inherent in the cell-like proteinoid structure derived. He makes such observations without suffering the confusion provided by the presence of other components of the contemporary cell. Lehninger (1975, p. 1045) has listed what, in his view, was needed in a primordial cell. The research indicates that many properties beyond those called for by Lehninger are present in proteinoid microsystems.

While it is axiomatic that proteinoid microsystems could not be converted into a contemporary cell until code-functional nucleic acids and efficient membrane lipids were acquired, both some molecular order and some selective diffusion through the boundaries are represented in the simplest proteinoid microsystem.

The degree of specialization in the contemporary cell is such that 1) order in polyamino acids and 2) barrier properties in membranes depend on different kinds of substance, e.g., nucleic acid and lipid, respectively (cf. Oparin, 1957, p. 287; Tien, 1974). To the extent that functions of the contemporary cell can be related to primordial properties in a simulated protocell, however, the more probable it is that such simulation is relevant. In this section, the proposed primordia of various current cellular functions are reviewed, as are those experiments that project the simulated proto-cell partway to the contemporary cell for each function. Some speculation on how the remaining gap may be bridged is included in some instances.

Ordered Microsystems

The existence of a high degree of order in the copolyamino acids that are aggregated into optically discernible microsystems, and some of the mechanism for that ordering, have been discussed above. The ease with which these microsystems appear and the regularity of their pattern are illustrated in Figures 6 and 8. The degree of order in the microsystems is regarded as possible because of the order in the component macromolecules.

As Samuel (1972) has pointed out, order is of many kinds and is expressed at many levels. The degree of order of the many individual microsystems is expressed especially in their usually uniform diameter. The order in the macromolecules (typically 10^{10} of them per microsystem) is conveyed to the supramolecular structure by what appear to be very strong intermacromolecular forces. The strong binding tendencies of proteinoid molecules is recognized by the tenacity with which these molecules adhere both to other large, and to small, molecules during various procedures of fractionation.

When compared with more highly evolved cells of some protists, the simulated protocells appear more ordered to the eye. An evolution of protocells to less highly ordered contemporary cells is suggested.

Although parent-connected multiplication of proteinoid microsystems could have persisted in an archaic phase extending over millions of years, and covering countless natural experiments, some of those experiments would have had to lead to internal synthesis of ordered polypeptides. The basis for this statement is simply that the type capable of internal synthesis represents what we know to be contemporaneous.

Since we view ordered morphology as a derivative of ordered macromolecules, the evolutionary maintenance and development of the order within the initial polyamino acid molecules is relevant. The research described in this chapter would not have been started if a way to ordered macromolecules had not been visualized.

In comparing molecular order in proteinoid microsystems with molecular order in contemporary cells, one criterion for which data are available is that of amino acid composition. The similarities in composition between proteinoids and proteins have occasionally been a subject for remarks, have been documented (Krampitz and Fox, 1969), and have been analyzed quantitatively (cf. Mikelsaar, 1975). As explained earlier, the similarities found between proteinoids and proteins can be rationalized by the explanation that bidirectional recognition between polyamino acids and polynucleotides modulated to the contemporary sequence of DNA → RNA →

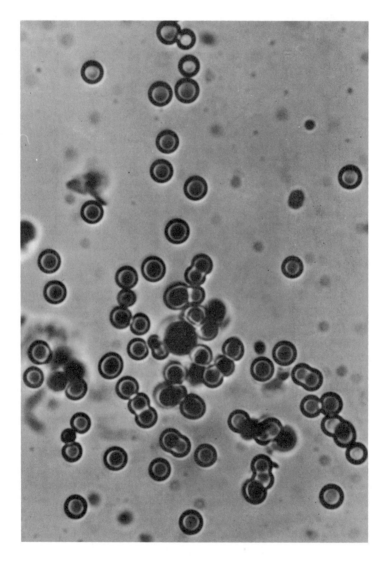

Figure 8. A field of proteinoid microsystems. × 300.

protein (Fox, 1959; Jukes, 1966; Fox and Dose, 1972; Chernavskii and Chernavskaya, 1975; Mikelsaar, 1975) through protoribosomes in a series of related steps. An alternative explanation for the documented similarities in composition of proteinoids and proteins is that similar forces were operating, in parallel (rather than in series), at the evolutionary stage of proteinoid formation and at the stage of nucleic acid–influenced protein biosynthesis.

Lipid Quality in Proteinoid Microsystems

Substantial fractions of a number of proteinoids are soluble in alcohol or in acetone. This suggests some lipid quality in the material. As Lehninger (1975, p. 1047) has pointed out, apolar amino acid sidechains in proteinoid (or protein for that matter) have lipid quality. Contributory functions like those of lipid are indicated by responses to atonic solutions (Fox et al., 1969); the microspheres shrink in hypertonic solution and swell in hypotonic solution. The microsystems also permit the passage of small molecules far more readily than they allow macromolecules to diffuse through the boundary (Fox et al., 1969). (In unpublished studies, Gilbert Baumann has found that he can make black films, from some proteinoids, in the Mueller-Rudin apparatus used to study lipid bilayers. The proteinoid films are self-sealing when ruptured, like lipid films. They thus clearly have lipid properties.)

Such membranes are more leaky than the contemporary ones that include in their makeup discrete lipids. We can visualize the development of a more efficient membrane during evolution, upon the cellular synthesis of lipids. Moreover, it seems reasonable that in the earliest stages of evolution of cellular metabolism free diffusion of small molecules into the protocells was helpful, or necessary, to provide the developing cells with some molecules as metabolic intermediates from the environment, until early cells could have evolved to make their own.

Another explanation that has been offered for the origin of lipids is found in a report that glycerol, lipids, and phosphate yielded phospholipids (Hargreaves, Mulvihill, and Deamer, 1977) when these were warmed under conditions that have been used for amino acids alone (Harada and Fox, 1965). While this route deserves to be considered, the most plausible origins are those that require a minimal assembly of intermediates or of products from various sites.

According to these experiments, the proteinoid microsystem

began with ordered molecules and a lipid-like barrier, as aspects of the proteinoid itself. At that stage, discrete lipid and nucleic acids were not needed. For evolution to the contemporary cell, internal syntheses of various molecules would have had to emerge next.

The Boundary of the Cell

The membrane of the contemporary cell enters into some cellular reactions, it serves as a selective barrier for molecules, and it contains the functioning cytoplasm. In this last role, it is the wrapping of the micropackage. Various functions of the membrane are listed in Table 6 (Fox, 1976b). Undoubtedly, functions and properties other than those listed in Table 6 should be included.

An especially significant benefit of the cell at all stages of evolution has been the protection of reactions by the provision of hydrophobic zones. These zones have separated the cellular chemistry from thermodynamically unfavorable concentrations of water while permitting the contiguity of aqueous zones.

The benefits listed apply qualitatively to both the contemporary cell and the protocell. They offer much individually, and in collective integration, to the cell as a biological quantum. Accordingly it can be inferred that no other line of evolution could have competed successfully with the cellular type. The cellular line of evolution was able to begin very early because, as the experiments demonstrate, an early kind of cell (Lehninger, 1975, p. 1045) could have arisen easily, often, and rapidly (Fox and Dose, 1977).

Table 6.   Benefits of a cellular membrane to evolution

Physical protection of organic material
Organization of chemical reactions
Compartmentalization of functions
Thermodynamically favorable hydrophobic zones
Maintenance of kinetically favorable concentrations
Reproduction at microsystemic level
Adaptive selection (Darwinian) of individual variants at microsystemic level
Screening of macromolecules from diffusible molecules
Promotion of dynamic interactions with the environment
Juxtaposition of enzymes, organelles, etc.
Systemic support and development of photosynthesis
Enlargement of metabolic pathways through (re)combination
Favorable spatial relations for coding interactions

Morphology

The above described awareness is one that emerged from experiments. In the absence of this new experimentally derived perspective, the seemingly logical assumption is a hierarchical progression (pp. 36–37 above), a progression from monomers through polymers to supramolecular systems. However, as stated, a minimal, reproductive cell composed of copolyamino acid alone would be a system in which both nucleic acids and true proteins could arise and evolve (Fox, 1965). This inference is consistent with an integrated overviewing continuum (Figure 2).

A view of this proteinoid microsystem under the electron microscope is presented in Figure 9. The multilaminar nature of the boundary is evident.

A scanning electron micrograph of a cluster of such microsystems is seen in Figure 10. Some junctions are visible in the photograph.

Comparisons of sections of bacteria with those of proteinoid microsystems are seen in the electron micrographs of Figure 11. Although the proteinoid figures are on the simple side, so are these bacteria, which are most likely to be lineal descendants of the most primitive types. Proteinoid tends to aggregate into spherical structures. Some proteinoids tend to form tubular structures, a propensity that is magnified by complexing with contemporary lipids.

In recent years, microfossils attributed to fungal and bacterial origin have been of general interest (Schopf, 1976). Many of these microfossils (Figure 12) are indistinguishable from proteinoid microsystems (Fox and Dose, 1977). Recently, Dr. Lynn Margulis has made artificial fossils of both fungi and proteinoid microsystems. The resultant observation is that all of the microfossils are open to serious reinterpretation (Francis, Margulis, and Barghoorn, 1978) as perhaps fossilized microspheres. If the suggestion is correct, paleontological proteinoid microspheres are at hand.

Osmotic Properties

The osmotic properties and the tendencies of molecules to diffuse through the membranes have been reviewed (Fox et al., 1969). Proteinoid membranes are relatively leaky (Stillwell, 1976) compared to most contemporary cellular membranes. However, as stated earlier, until the early evolving cell learned to make nearly all of its own metabolic intermediates, it should have benefited from permitting a variety of small molecules to diffuse inward

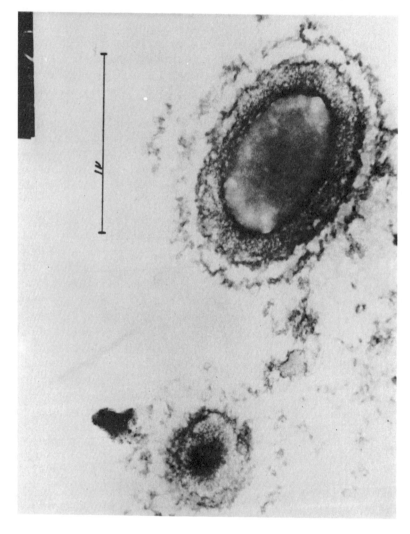

Figure 9. Electron micrograph of proteinoid microsystem. (Courtesy of Dr. Walther Stoeckenius.)

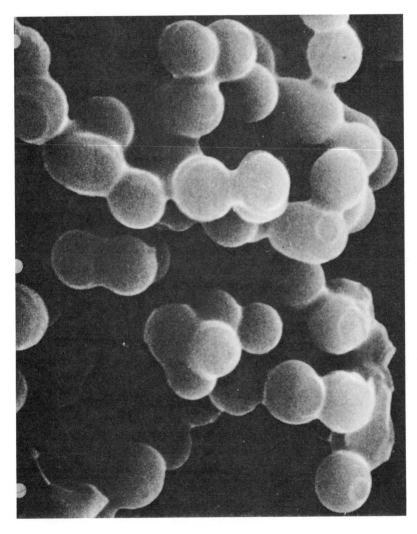

Figure 10. Scanning electron micrograph of microsystems composed of proline-rich proteinoid. (Courtesy of Mr. Steven Brooke.)

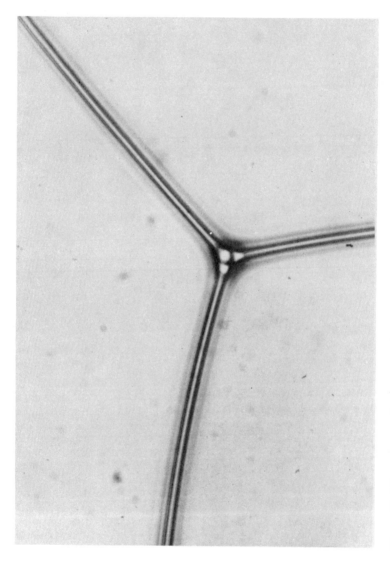

Figure 11. Microstructures from proteinoid and α-lecithin. (Courtesy of Dr. Y. Yamagata.)

from the environment through the membrane. As the cell evolved a complete, increasingly contemporary metabolism, it could afford, and benefitted from, development of the ability to retain continually smaller molecules (Fox, 1977).

Compartmentalization and the Gram Stain

Our assumption, again, is that metabolism and the cell evolved together from their simplest beginnings. The cell that we know today has a number of compartments. One part of the cell is able to conduct one kind of biochemical business, or to do it at one pH, while another part conducts reactions differently. Systematic experiments have indicated that the simulated protocell is easily subject to compartmentalization (Figure 13). This principle of construction, thus, did not need to wait on evolution from the protocell.

Many kinds of compartmentalization have been shown in a number of ways (Brooke and Fox, 1977); calcium is especially contributory. A vivid verification of compartmentalization is attained by the Gram stain. In a single microsphere, some parts may stain Gram-positive while others stain Gram-negative (Figure 14).

The studies of compartmentalization have revealed a maximally simple example of the components of a protoselection, or preselection, process. When proteinoid microspheres are compartmentalized, some undergo more compartmentalization than others. The more ramified ones display greater durability as the pH is raised, whereas the less ramified particles are more immediately subject to extinction (Brooke and Fox, 1977). The surviving individuals do not, however, bequeath their durability to "offspring." Such a stage in selection would have had to wait on a stage in which microspheres could themselves synthesize material from which offspring primitive cells could form.

Proliferation

In the contemporary cell, reproduction occurs at two levels: the molecular and the cellular. The instructions for making the components of the cell are encoded in the DNA; the identity of the cells in the next generation is thus specified by the DNA. Within the span of one cycle of the cell, the molecule thus appears to precede the supramolecular system.

It does not follow that the first process of reproduction on

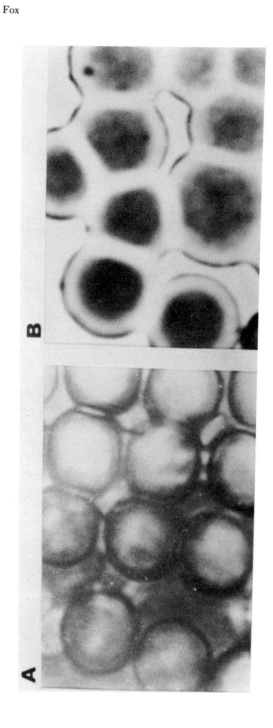

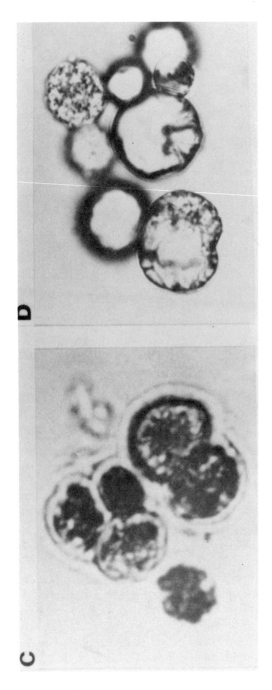

Figure 12. Comparison of microfossils with microsystems (Fox and Dose, 1977). Microfossils in left column, proteinoid microspheres in right column.

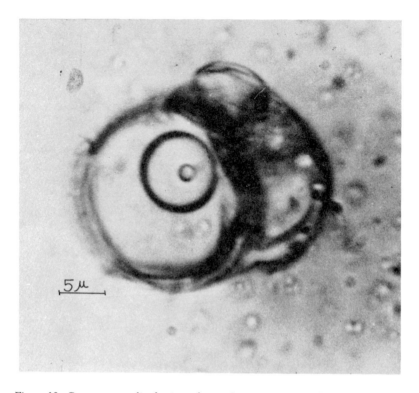

Figure 13. Compartmentalized microspheres. Concentric microspheres may also be seen. (From S. Brooke, M.S. thesis, University of Miami, 1976.)

Earth was an expression of prior DNA on Earth. DNA as we know it replicates within a cell. That replication is made possible by polyamino acids such as enzymes. A relevant sequence of events is thus the origin of polyamino acids and the origin of proliferative protocells therefrom, with DNA to have come later. The sequence of amino acids → polyamino acid → simulated protocell has been demonstrated by experiments, as well as the ability of such simulated protocells to catalyze the formation of internucleotide bonds (Jungck and Fox, 1973).

For the alternative concept of DNA-first (Crick, 1968), no experimental demonstration has been reported. Nor has the production of any functionally DNA-like polymer been reported ab initio from abiotic materials.

Component processes of cycles of parent-dependent multiplication of simulated protocells have been described. These have

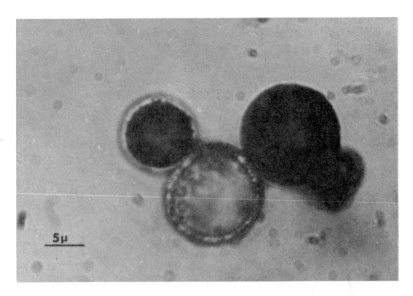

Figure 14. Gram-positive and Gram-negative staining in the same microsphere. (From S. Brooke, M.S. thesis, University of Miami, 1976.)

included prototypal equivalents of budding (Fox, McCauley, and Wood, 1967, Figure 15), binary fission (Figure 16), sporulation (Figure 17), and parturition (Figure 18). In each of these cases, growth has occurred only by presumably primitive processes of accretion. In none of these systems has any polynucleotide been present, or needed.

As already demonstrated, the material of the simulated proto-cells has a high degree of molecular order, and it is chemically reactive and photoreactive. Such order would influence order in macromolecules that might be first manufactured within proto-cells. The membranes would constitute a reactive zone in the units.

Worthy of note is the fact that molecular ordering could have occurred first in the absence of cell-like microsystems, *whereas* the *proliferation* of the derived microsystems *would have occurred at a higher level* and a later time.

Motility

Proteinoid microparticles are of a size to undergo Brownian motion. They have often been observed to perform in this way. When the microspheres are composed of a zinc salt of proteinoid and ATP is

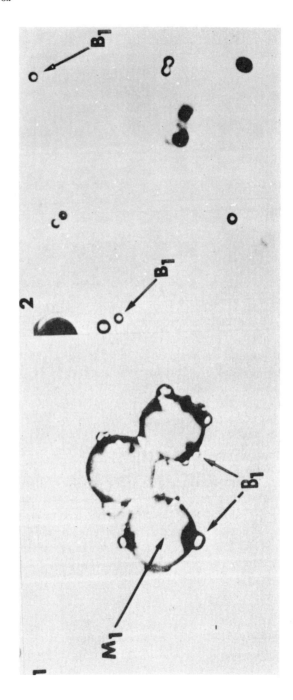

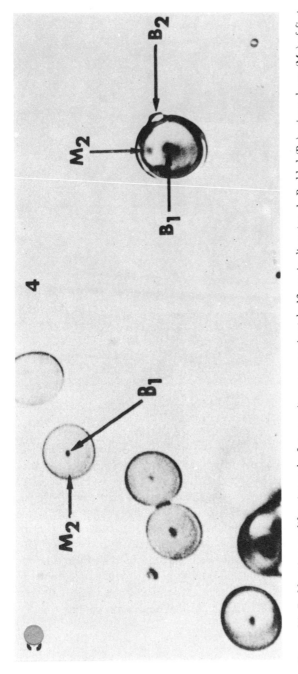

Figure 15. Budding in a proliferative cycle. Large units are approximately 12 μm in diameter. 1. Budded (B_1) microspheres (M_1) of first generation. 2. Separated buds. 3. Microspheres (M_2) grown from central, darkly stained buds. 4. Mature second-generation microsphere (M_2) with bud (B_2).

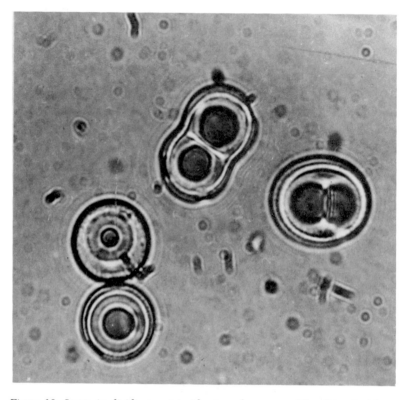

Figure 16. Stages in dividing proteinoid microspheres resembling those in binary fission. In this case, prior coalescence of two units has not been ruled out.

added to the suspension, the motion is especially vigorous. In the course of such experiments an asymmetrical particle was photographed under the microscope. Like other asymmetrical particles, it tumbled and moved about in a manner that, to the eye, suggested nonrandom motion (Figure 19). Nonexpert student observers have characterized this behavior as purposeful. Indeed, ostensibly "purposeful" behavior may have evolved, by selection, from such physical response to stimuli. When one extrapolates from this behavior, from that of contemporary organisms, and from the activities of Adler's bacteria (Adler, 1969), a potential evolutionary concept for motility and its utility begins to emerge.

Formation of Junctions and Aging
One of the more regular phenomena revealed by simple observation

of the proteinoid microsystems is that of the ease with which they form functions as a consequence of the motion (Hsu, Brooke, and Fox, 1971). The tendency to form such junctions is illustrated in Figure 19. The reality of such junctions is supported by Figure 20, which shows fracture of junctions and jagged separation of two joined microspheres. Observation reveals the freshly prepared microspheres to have very fluid membranes. On standing, within a period of, typically, three days, the junctions stiffen. Some kind of aging is evidently occurring.

When water is allowed to flow through a cluster of joined microspheres, the interiors dissipate to small endoparticles, while the boundaries persist. When the endoparticles become smaller than the junctions, they frequently pass through those junctions (Figure 21). This indicates that the junctions are mostly, or entirely, hollow.

The phenomenon also presents us with a model for elementary communication; the proteinoid endoparticles are informational (see footnote 9) in that proteinoids interact selectively with each other (Hsu and Fox, 1976) and with enzyme substrates (Fox and Dose, 1977), and the basic proteinoids have been shown to interact selectively with polynucleotides (Yuki and Fox, 1969).

CHEMICAL ACTIVITIES IN PROTEINOID MICROSYSTEMS

An ongoing interest in the study of proteinoid microsystems relates to whether or not proteinoid microsystems can evolve and, more specifically, whether they can evolve metabolically. Theoretically, we can anticipate that each of the catalytic activities demonstrated in proteinoids (Table 3) would at first be carried into the microsystems assembled from such proteinoids.

A factual demonstration of such transfer from the macromolecular to the microsystemic level has been accomplished in some studies. In the first of these, the so-called 2:2:1-proteinoid catalyzed the conversion of glucose to glucuronic acid and the decarboxylation of the latter (Fox and Krampitz, 1964). Microspheres in suspension showed about the same degree of activity (Table 7). It is of interest that basic proteinoids are more powerful catalysts in the breakdown of glucose than are acidic polymers. However, the acidic ones are convertible to microsystems in much greater abundance. This experiment deserves to be repeated with microsystems composed from a mixture of acidic and basic proteinoids.

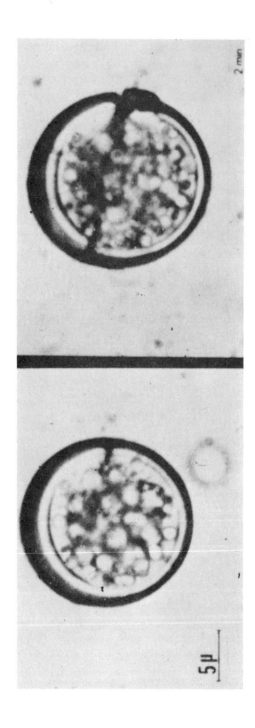

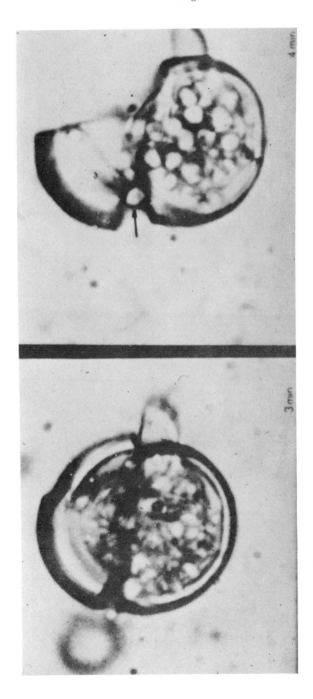

Figure 17. Simulated protosporulation from particles of acidic and basic proteinoids and calcium ion. 0, 2, 3, 4 min.

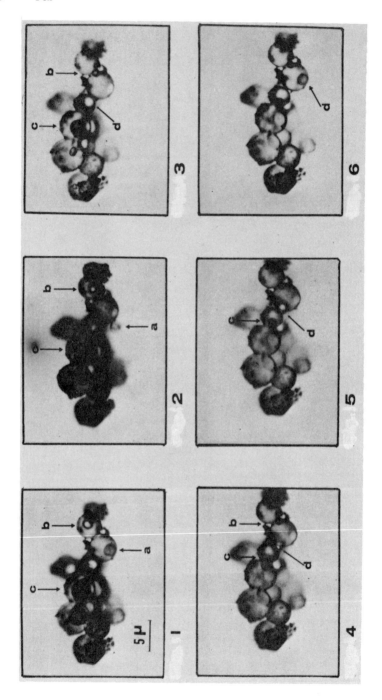

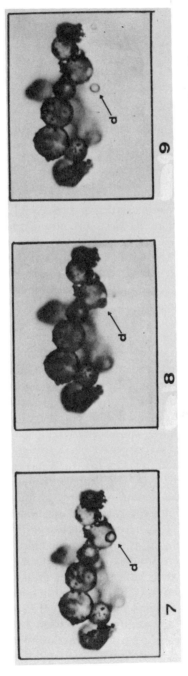

Figure 18. Simulation of primitive parturition. Released endoparticle is seen clearly in 9. It can grow by accretion as in Figure 15.

Table 7.   Evolution of CO_2 from radioglucose in incubation with proteinoids

Proteinoid	Radio CO_2 cpm
DJ acidic polymer	117
RM acidic polymer	62
Free amino acids of polymer	0–5
Hydrolyzate of DJ polymer	0–5
Acidic proteinoid microspheres	64

Background of 42–55 cpm. (From Fox and Krampitz, 1964.)

The known esterase activities of proteinoids (Fox and Dose, 1977) have been demonstrated in proteinoid microsystems. This was done at pH 4, which has been shown to allow only insignificant quantities of the proteinoid into solution from the phase-separated particles (Hsu and Fox, 1976).

Another proteinoid activity transferrable to microsystems is that of peroxidatic activity, demonstrated in hemoproteinoids by Dose and Zaki (1971).

Also shown has been the ease of combining esterase and peroxidase activity in the same microsystems (Hsu and Fox, 1976). Unexpectedly, when peroxidatic proteinoid was used to coat esterolytic proteinoid seeds, the peroxidase activity was enhanced over that observed in microsystems of peroxidatic proteinoid microsystems alone. The inversely prepared systems, peroxidatic seeds coated with esterolytic proteinoid, showed no synergistic effects.

Loss in catalytic activity can be expected when some proteinoids are aggregated by heating in aqueous solution. Some activities, such as esterolytic (Rohlfing and Fox, 1967) and phosphatatic (Oshima, 1968) activities, are lost or diminished by heating in the presence of solvent water. Others are enhanced by that treatment (Dose and Zaki, 1971). To avoid effects of heating in aqueous solution, acidic proteinoid and basic proteinoid can be combined in the cold. Such microspheres have not been studied much for catalytic activity. However, they have been examined for the ability to catalyze the production of internucleotide bonds. Jungck (1973) found that basic proteinoids would catalyze the formation of pentamers of adenylic acid in an anhydrous solvent. In aqueous solution the same proteinoid catalyzed the formation of adenine dinucleotide. When the basic proteinoid was converted to microspheres with

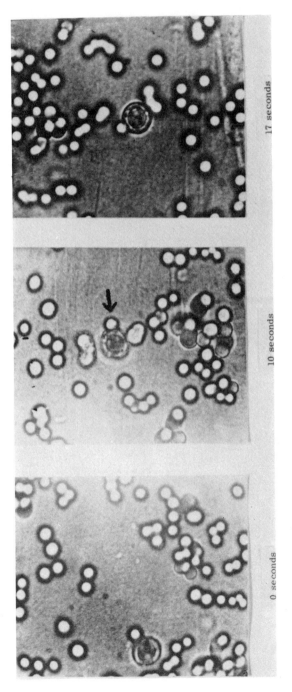

Figure 19. (Continued pages 74–75.)

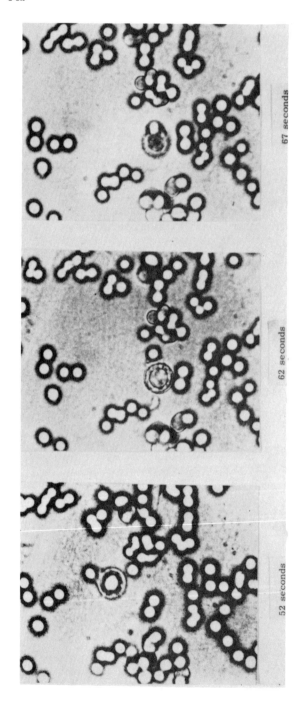

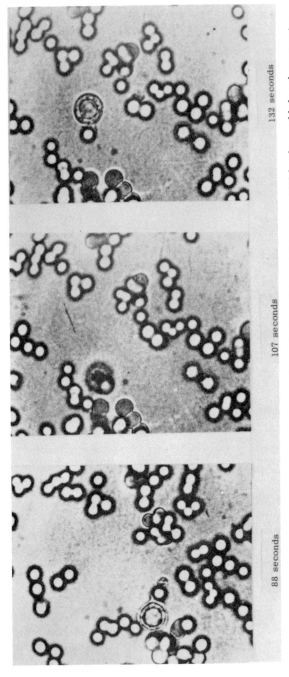

88 seconds   107 seconds   132 seconds

Figure 19. Apparently nonrandom motion in an asymmetrical proteinoid particle containing zinc. ATP has been added to the suspension. The asymmetric particle both moves and tumbles.

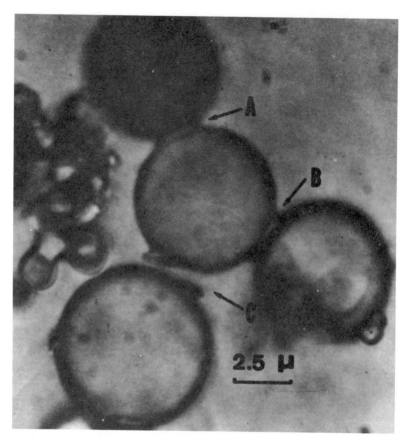

Figure 20. Evidence for junctions between proteinoid microspheres. The junctions have aged and stiffened, and have been stained. Intact (A), cracked (B), and separated (C) junctions are seen.

acidic proteinoid, Jungck recorded the formation of adenine tri-nucleotide as well as the dinucleotide (Jungck and Fox, 1973).

The production of larger polynucleotides in protocells under primitive conditions is a principal gap in the predicated primordial sequence. The possibility of obtaining a polymer larger than a trinucleotide is being investigated. Also being investigated is the hypothesis that the size of polynucleotides would grow while poly-amino acids grow (cf. Krampitz and Fox, 1969) during replicative generations of microsystems. A kind of reflexive escalation in molecular size might result; larger polynucleotides might result from larger polyamino acids and vice versa.

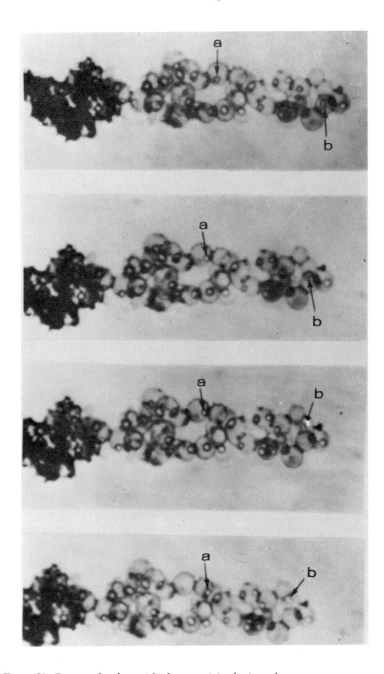

Figure 21. Passage of endoparticles between joined microspheres.

A most fundamental clarification of the evolvable metabolism in proteinoid microsystems would result from the demonstration of conversion of solar energy to chemical energy in microsystems. Logically, this would require the conversion of inorganic phosphate by light to inorganic pyrophosphate (Baltscheffsky, 1974) or to ATP from AMP or ADP, the origin of the last having been explained (Waehneldt and Fox, 1968; Ryan and Fox, 1973). Once an energy-rich material existed in early cells, comprehensive metabolism and its evolution could have flowed therefrom.

For this possibility, a body of unpublished information exists. These data include numerous indications of ATP synthesis from AMP and numerous striking indications of ATP instability under varied conditions. Starting with ADP, there are indications of ATP synthesis by white light, mostly in a nonaqueous solvent (dimethylformamide). Fan, Chien, and Chiang (1976) have indicated what may be a minute yield of ATP by photoillumination of suspensions of ADP. Most encouraging for the protobiogenecist is the synthesis of ATP in reconstituted purple membrane vesicles (Racker and Stoeckenius, 1974; Yoshida et al., 1975). Racker and Stoeckenius used ATPase in a reconstituted purple membrane system. The Yoshida group reports obtaining from ADP a 15% yield of ATP by illumination with a Kodak projector. No respiratory chain components were necessary. All of the components of the Yoshida system seem to be readily characterizable.

Because the components of the Yoshida system are mainly eight "subunits" of ATPase, we believe that thermal polyamino acids can be screened for the same kind of activity, inasmuch as so many enzyme-like and some photocatalytic activities (Fox and Dose, 1977) have been found in thermal polyamino acids.

ASSESSMENT OF THE GAP BETWEEN THE SIMULATED PROTOCELL AND THE CONTEMPORARY CELL

What is needed to bridge the gap between a protoorganism and a contemporary cell has been discussed previously (cf. Fox, 1975b). Some of the attempts to bridge the gap and some of what needs to be done in the future have been considered in this chapter.

The present outlook on the gap, as derived from constructionistic experiments, recognizes that the first organisms might have been photochemically active, and might therefore have had a significant measure of autotrophism (Fox, 1974a; Hartman, 1975).

Although questions of the gap have been formulated from the contemporary cell (e.g., Horowitz, 1945; Granick, 1965; Lipmann, 1965; Florkin, 1975), the ultimate answers will have to interdigitate both with a mature theory of prebiotic evolution ab initio and with the disciplined knowledge of the contemporary. To date, some of the principles of primordial assembly in evolution in the forward direction have been deduced only from heuristic experiments (Fox, 1975c). These tenets (Fox, 1977) include, for example, the principle of self-ordering of amino acids, and an explanation of how bidirectional recognition between amino acids and nucleotides became the sequence stated by the Central Dogma of molecular biology (Fox, 1975b).

Experiments have begun to indicate how some internucleotide bonds might have formed in protocellular metabolism. Models for the formation of protoribosomes and some slight suggestion for the formation of phenylalanine peptides (Fox, Jungck, and Nakashima, 1974) exist. Experiments under way, based on those of Weber et al. (1977), extend understanding of peptide synthesis in concentrated aqueous solution, a condition that probably describes the surface of ribosomes and the surface of protoribosomes. Experiments concerning interaction of copolyamino acids with homopolyribonucleotides indicate that the affinities of the genetic code are intrinsic to the interacting molecules. We may just be beginning to understand how the evolution of protocellular systems was adapted to light energy, an adaptation that was a punctuation point for all organic evolution (Table 8 and Figure 22).

Both randomness and hierarchical progression are concepts that have been rather widely held by those who seek to infer the nature of the origin of life by analysis. Despite the seemingly logical nature of those concepts, they are not supported by experimental results on the production of macromolecules and phase-separated particles, and they have not been interpolated into an integrated

Table 8. Some functions needed to convert the proteinoid microsphere to a more contemporary cell

---

Synthesis of cellular nucleic acid
Synthesis of cellular protein
A genetic code relationship between nucleic acids and cellular proteins
Mechanism for conversion of solar energy to chemical energy, directly or
    indirectly

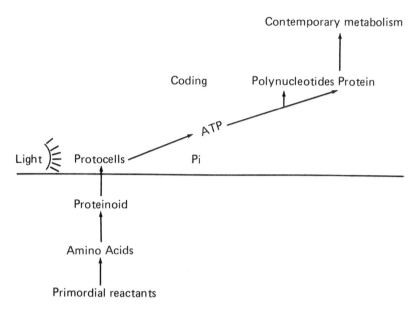

Figure 22. The current form of the flowsheet. In this flowsheet, the beginning of reproduction was intrinsic to the protocells. Molecular reproduction originated as a facet of contemporary metabolism.

theory; nor, after early attempts at producing prebiotic nucleic acids (Schwartz, Bradley, and Fox, 1965), do I see how they can be. Most unexpected were the ease and efficiency with which both the origin of minimal cells and their parent-connected multiplication (protoreproduction) occurred experimentally by aggregation of thermal proteinoid. Conceptually, the protoreproduction does not require prior protoreproduction at the molecular level.

What was needed in addition to analysis of the contemporary cell was geochemical reasoning and heuristic experimentation ab initio from amino acids. The results of individual experiments then dictated the design of subsequent experiments, finally to suggest a narrow, internally directed evolutionary pathway. For this pathway, the principle of self-ordering of amino acids was crucial.

CRITERIA OF PLAUSIBILITY AND VALIDITY

Frequently the statement is made that the protobiogeneticist can do no more than attempt to simulate events of three billion years ago; he can learn only what could happen, not what did happen. Each new area of science seems, however, to develop its own criteria of validity. Other theories are also distant from the occurrence, not necessarily in time. For example, the atomic theory is distant from the viewer in dimension. He has not been able to see the atoms.

One criterion for validity of the theory of protobiogenesis is geological relevance of the experiments. Another criterion is that of sequential compatibility of the steps; can one stage serve as a matrix for the next? The relevance of this criterion is indeed a functional feedback of the theory itself because its strength depends on the view that each stage is matrix-dependent as in true evolution.

Each stage is subject to overlapping criticism from biological, chemical, physical, and geological perspectives. Any inference that meets all of these superposed perspectives has survived several challenges simultaneously.

Basically, the candidate events for protobiogenesis are considered in the context of a sequence of events and are quantitatively to be judged as more or less probable. With the multiplication of probabilities of individual steps, quantitative differences, i.e., various degrees of plausibility for individual steps, can become in effect qualitative differences, i.e., validities versus invalidities.

Plausibility of a concept, model, or theory requires, by definition, concurrent acceptability by a number of students of a given thesis. For this physical model and the derived theory, representative students, for example, of a number of disciplines have commented to various degrees and in various ways, under the headings of nonrandom composition or sequence (Ross, 1962; Calvin, 1969; Ambrose and Easty, 1970; Baker and Allen, 1971; Broda, 1975; Dickerson and Geis, 1976), enzymic activity (Knight, 1974; Broda, 1975; Florkin, 1975; Lehninger, 1975; Dickerson and Geis, 1976; Weinberg, 1977), morphological resemblance to cells (Oparin, 1968; Glasstone, 1968; Dowben, 1971; Fuller, 1974; Knight, 1974; Price, 1974; Turcotte, Nordmann, and Cisne, 1974; Wolken, 1975; Mader, 1976; Volpe, 1977; Weinberg, 1977), dynamic resemblance to cells including ability to "reproduce" (Ambrose and Easty, 1970; CRM, 1972; Kenyon, 1973; Fuller, 1974; Florkin, 1975; Lehninger, 1975; Dickerson and Geis, 1976; Winchester, 1976; Hartl, 1977; Weinberg, 1977; Winchester, 1977; Wolfe, 1977), and evolvability in general (Young, 1966; Baker and Allen, 1971; Korn and Korn, 1971; Vegotsky, 1972; Brooks and Shaw, 1973; Easton, 1973; Kenyon, 1974; Knight, 1974; Dickerson and Geis, 1976; Winchester, 1976).

The fullest plausibility for any simulated protoorganism, however, requires integrated plausibilities joined with a plausible source.

Some theories have originally attained a state of acceptance by providing a satisfactory explanation relating to large bodies of information and by continued pervasiveness despite extensive and varied challenges. One example of such a principle is the second law of thermodynamics. Many attempts to disprove the second law "have served rather to convince the incredulous and to establish the second law of thermodynamics as one of the foundations of modern science" (Lewis and Randall, 1961). As is well known, Darwin's theory of selection codifies large bodies of systematic biological information. Like the second law, it conflicted with prevailing views, so that its period of testing was prolonged. Its base of strength, as T. H. Huxley pointed out, was the nonexistence of a valid comprehensive, scientific alternative.

Although the proteinoid theory of protobiogenesis has its own characteristics and has not been extant for periods as long as several decades, it has features in common with the two other theories mentioned. It does relate to large bodies of information, and it has received challenges on specific issues, which have been answered. This theory has a special aspect of approachability in that a physical product is obtained. The product has increasingly been subject both to comparison with contemporary cells and to experimental modification.

An assessment of the validity of the proteinoid model and theory is dependent upon the validity of the answer for the chicken-egg question. If one assumes that functional nucleic acids arose before cells and proteins, the full proteinoid theory presented is not relevant to a theory of spontaneous protobiogenesis. I have not seen that the conceptual possibility of nucleic acids–first is supported by relevant experiments, nor have I seen any attempt to make that predicate conceptually defensible in a detailed analysis (cf. Florkin, 1975, p. 252). *Without some such support, the assumptions would leave us again with the unsolved chicken-egg problem.*

Some of the evolutionary sequence following the step of informational preprotein → protocell in the proteinoid theory has however been simulated in the laboratory, and that step is based on the demonstrated phenomenon of self-ordering of amino acids. This resolution of the chicken-egg question permits, as discussed earlier in this chapter, a stepwise evolution up to a protoreproductive protocell. No insurmountable conceptual barriers are seen beyond that stage (cf. Ehrensvärd, 1962) although many questions are yet to be answered for the subsequent steps.

The principle of the self-ordering of amino acids is perhaps the most fundamental new knowledge to arise from constructionistic studies (Fox, 1977); it did not emerge from related analytical studies (Fox, 1976c). It is based on a demonstrated phenomenon for which the mechanism is now partly understood.

The experiments with molecules and macromolecules and with supramolecular systems produced from those macromolecules in a continuum emphasize a nonrandom evolution to the protocell.

EPILOGUE

The summary of data, the interpretations, the flowsheet, and the explanatory mechanisms presented here may be reviewed in a historical context. The questions treated are related to that put by L. Pasteur in 1864 and again quoted in this chapter, "Can matter organize itself? In other words, are there beings that can come into the world without parents, without ancestors? That is the question to be resolved."

The history of the subject matter of protolife has centered about Pasteur's question and a few other key questions. They have been, principally, 1) the origin of the protoorganism (Pasteur's "beings"), 2) the chicken-egg questions for prebiomacromolecular times and for the earliest nucleoprotein stage of evolution, and 3) some newer questions. The primary chicken-egg question could not be asked in Pasteur's time because the chemistry of living systems was not sufficiently advanced to recognize nucleic acid and only minimally at a stage of apprehending protein.

Questions 1 and 2 listed above relate to the origin of the protoorganism. The nature of the protocell is a question that could be put after answers to these first two questions on origin were obtained. Numerous properties have been catalogued. A relatively comprehensive theory has emerged. A number of a priori hypotheses have been shown to be wanting, e.g., the hypothesis of unqualified hierarchical progression.

We can say, in response to Pasteur's question and in his context, "Yes, matter can organize itself." We can provide also at least a first answer to other questions. The kind of matter appears to be a heteropoly-α-amino acid formed by the heating of amino acids, the latter presumably available in sets on the primitive Earth through hydrolysis of precursors. The amino acids evidently contained the instructions for their own sequential arrangements in polymers. A number of laboratories have described many properties in the resultant polymers. The ways in which such polymers assemble have been defined. Many properties of products of such aggregation, simulated protocells, have been catalogued. This catalog has permitted a definition of the gap between a protocell and a contemporary cell.

The question of the origin of the first cells, glyoxal (Groth and Suess, 1938), etc., has many interfaces, is of widening interest, and is increasingly cited as relating meaningfully to such contemporaneously societal issues as cancer (Szent-Györgyi, 1977) and the material basis for the mind (Delbrück, Chapter 5, this volume). We may not only intellectually want, but physiologically need, to answer the remaining questions of the gap.

The next questions are indeed those of how the protoorganism could have become a contemporary organism. Analysis of the problems of the gap has become possible, and some experimental bridging of the gap has occurred. The fact that answers up to and at least a little beyond the protoorganism are possible encourages us to believe that we may eventually understand the nature of life in the powerful light of the origin and nature of protolife.

ACKNOWLEDGMENTS

The author acknowledges the contributions from a number of dedicated associates of experimental skill and help in keeping concepts close to the facts. Many of these associates are identified in the publications that have appeared in scientific journals. Thanks are extended especially to the National Aeronautics and Space Administration for fostering a continuous research effort without frequent interruptions for documentation other than scientific papers.

REFERENCES

Abelson, P. H. 1957. Discussion. Ann. N.Y. Acad. Sci. 69:274.
Adler, J. 1969. Chemoreceptors in bacteria. Science 166:1588–1597.
American Institute of Biological Sciences. 1963. Biological Science: Molecules to Man. Houghton Mifflin, Boston.
Ambrose, E. J., and Easty, D. M. 1970. Cell Biology. Addison-Wesley Publishing Co., Reading, Mass. p. 479.

Ashby, W. R. 1960. The self-reproducing system. In C. A. Muses (ed.), Aspects of the Theory of Artificial Intelligence, pp. 9–18. Proc. First Intl. Symp. on Biosimulation, Locarno.

Aw, S. E. 1976. Chemical Evolution. University Education Press, Singapore. p. 148.

Bagnara, J. T., and Hadley, M. E. 1970. Intermedin (MSH) {melanophore-stimulating hormone}-like effect of a thermal polymer on vertebrate chromatophores. Experientia 26:167–169.

Baker, J. J. W., and Allen, G. E. 1971. The Study of Biology. 2nd Ed. Addison-Wesley Publishing Co., Reading, Mass. pp. 792–795.

Baltscheffsky, H. 1974. Protein structure and the molecular evolution of biological energy conversion. In K. Dose, S. W. Fox, G. A. Deborin, and T. E. Pavlovskaya (eds.), The Origin of Life and Evolutionary Biochemistry, pp. 9–19. Plenum Press, New York.

Blum, H. F. 1955. Time's Arrow and Evolution. Princeton University Press, Princeton, N.J.

Breger, I. A., Zubovic, P., Chandler, J. C., and Clarke, R. S. 1972. Occurrence and significance of formaldehyde in the Allende carbonaceous chondrite. Nature 236:155–158.

Broda, E. 1975. The Evolution of the Bioenergetic Processes. Pergamon Press, Oxford, England.

Brooke, S., and Fox, S. W. 1977. Compartmentalization in proteinoid microsystems. BioSystems 9:1–22.

Brooks, J., and Shaw, G. 1973. Origin and Development of Living Systems. Academic Press, London.

Calvin, M. 1962. The origin of life on Earth and elsewhere. Perspect. Biol. Med. 5:413.

Calvin, M. 1969. Chemical Evolution. Oxford University Press, London.

Carter, C. W., Jr., and Kraut, J. 1974. A proposed model for interaction of polypeptides with RNA. Proc. Natl. Acad. Sci. USA 71:283–287.

Chernavskii, D. S., and Chernavskaya, N. M. 1975. Some theoretical aspects of the problem of life origin. J. Theor. Biol. 50:13–23.

Copeland, J. J. 1936. Yellowstone thermal Myxophyceae, Ann. N.Y. Acad. Sci. 36:1–232.

Crick, F. H. C. 1968. The origin of the genetic code. J. Mol. Biol. 38:367–379.

CRM. 1972. Biology, an Appreciation of Life. Del Mar, Cal. pp. 51–52.

Dickerson, R. E., and Geis, I. 1976. Chemistry, Matter and the Universe. W. A. Benjamin, Inc., Menlo Park, Cal. pp. 641–642.

Dose, K., and Rauchfuss, H. 1972. On the electrophoretic behavior of thermal polymers of amino acids. In D. L. Rohlfing and A. I. Oparin (eds.), Molecular Evolution: Prebiological and Biological, pp. 199–217. Plenum Press, New York.

Dose, K., and Zaki, 1971. Hamoproteinoide mit peroxidatischer und katalatischer Aktivitat. Z. Naturforsch. 26b:144–148.

Dowben, R. M. 1971. Cell Biology. Harper and Row, New York. p. 530.

Easton, T. A. 1973. A note on the mathematics of microsphere division, Bull. Math. Biol. 35:259–262.

Ehrensvärd, G. 1962. Life: Origin and Development. (Eng. trans.) University of Chicago Press, Chicago.

Eigen, M. 1971. Self-organization of matter and the evolution of biological macromolecules. Naturwissenschaften 58:465–523.

Fan, I.-J., Chien, Y.-C., and Chiang, I.-H. 1976. Inorganic photophosphorylation of adenosine diphosphate to adenosine triphosphate, Sci. Sin. 19:805–810.

Florkin, M. 1975. Comprehensive Biochemistry. Vol. 29B. Elsevier, Amsterdam.

Fox, S. W. 1957. The chemical problem of spontaneous generation, J. Chem. Educ. 34:472–479.

Fox, S. W. 1959. Biological overtones of the thermal theory of biochemical origins. Bull. Am. Inst. Biol. Sci. 9:20–23.

Fox, S. W. 1960. How did life begin? Science 132:200–208.

Fox, S. W. 1965. Experiments suggesting evolution to proteins. In V. Bryson and H. J. Vogel (eds.), Evolving Genes and Proteins, pp. 359–369. Academic Press, New York.

Fox, S. W. 1967. (Untitled remarks.) In P. S. Moorhead and M. M. Kaplan (eds.), Mathematical Challenges to the Neo-Darwinian Interpretation of Evolution. Wistar Institute Press, Philadelphia.

Fox, S. W. 1968. Spontaneous generation, the origin of life, and self assembly. Curr. Mod. Biol. (BioSystems) 2:235–240.

Fox, S. W. 1971. Chemical origins of cells—2. Chem. Eng. News 49(50): 46–53.

Fox, S. W. 1973a. The Apollo program and amino acids. Bull. Atom. Sci. 29(10):46–51.

Fox, S. W. 1973b. Origin of the cell: Experiments and premises. Naturwissenschaften 60:359–368.

Fox, S. W. 1973c. Molecular evolution to the first cells. Pure Appl. Chem. 34:641–669.

Fox, S. W. 1974a. Origins of biological information and the genetic code. Mol. Cell. Biochem. 3:129–142.

Fox, S. W. 1974b. Coacervate droplets, proteinoid microspheres, and the genetic apparatus. In K. Dose, S. W. Fox, T. E. Pavlovskaya, and G. A. Deborin (eds.), The Origin of Life and Evolutionary Biochemistry, pp. 119–132. Plenum Press, New York.

Fox, S. W. 1975a. Stereomolecular interactions and microsystems in experimental protobiogenesis. BioSystems 7:22–36.

Fox, S. W. 1975b. The matrix for the protobiological quantum: Cosmic casino or shapes of molecules? Intl. J. Quantum Chem. QBS2:307–320.

Fox, S. W. 1975c. Looking forward to the present. BioSystems 6:165–175.

Fox, S. W. 1976a. Response to comments on thermal polypeptides. J. Mol. Evol. 8:301–304.

Fox, S. W. 1976b. The evolutionary significance of phase-separated microsystems. Orig. Life 7:49–68.

Fox, S. W. 1976c. The evolutionary significance of ordering phenomena in thermal proteinoids and proteins. In J. L. Fox, Z. Deyl, and A. Blazej (eds.), Protein Structure and Evolution, pp. 125–148. Marcel Dekker, New York.

Fox, S. W. 1977. Bioorganic chemistry and the emergence of the first cell. In E. E. van Tamelen (ed.), Bioorganic Chemistry, Vol. III, pp. 21–32. Academic Press, New York.

Fox, S. W., and Dose, K. 1972. Molecular Evolution and the Origin of Life. Freeman, San Francisco.

Fox, S. W., and Dose, K. 1977. Molecular Evolution and the Origin of Life. Rev. ed. Marcel Dekker, New York.

Fox, S. W., Harada, K., and Hare, P. E. 1976. Amino acid precursors in lunar fines; limits to the contribution of jet exhaust. Geochim. Cosmochim. Acta 40:1069–1071.

Fox, S. W., Harada, K., and Rohlfing, D. L. 1962. The thermal copolymerization of α-amino acids. In M. A. Stahmann (ed.), Polyamino Acids, Polypeptides, and Proteins, pp. 47–53. University of Wisconsin Press, Madison.

Fox, S. W., Harada, K., and Vegotsky, A. 1959. Thermal polymerization of amino acids and a theory of biochemical origins. Experientia 15:81–84.

Fox, S. W., Jungck, J. R., and Nakashima, T. 1974. From proteinoid microsphere to contemporary cell: Formation of internucleotide and peptide bonds by proteinoid particles. Orig. Life 5:227–237.

Fox, S. W., and Krampitz, G. 1964. The catalytic decomposition of glucose in aqueous solution by thermal proteinoids. Nature 203:1362–1364.

Fox, S. W., Lacey, J. C., Jr. and Nakashima, T. 1971. Interactions of thermal proteinoids with polynucleotides. In D. Ribbons and F. Woessner (eds.), Nucleic Acid–Protein Interactions, pp. 113–127. North-Holland Publishing Co., Amsterdam.

Fox, S. W., and McCauley, R. J. 1968. Could life originate now? J. Am. Museum Natural Hist. 77(7):26–31.

Fox, S. W., McCauley, R. J., Montgomery, P.O.'B., Fukushima, T., Harada, K., and Windsor, C. R. 1969. Membrane-like properties in microsystems assembled from synthetic protein-like polymer. In F. Snell, J. Wolken, G. J. Iverson, and J. Lam (eds.), Physical Principles of Biological Membranes, pp. 417–430. Gordon and Breach, New York.

Fox, S. W., McCauley, R. J., and Wood, A. 1967. A model of primitive heterotrophic proliferation. Comp. Biochem. Physiol. 20:773–778.

Fox, S. W., and Waehneldt, T. V. 1968. The thermal synthesis of neutral and basic proteinoids. Biochim. Biophys. Acta 160:246–249.

Fox, S. W., and Wang, C.-T. 1968. Melanocyte-stimulating hormone: Activity in thermal polymers of alpha-amino acids. Science 160:547–548.

Fox, S. W., Wiggert, E., and Joseph, D. 1965. Described in S. W. Fox (ed.), The Origins of Prebiological Systems, pp. 368–371. Academic Press, New York.

Fox, S. W., Winitz, M., and Pettinga, C. W. 1953. Enzymic synthesis of peptide bonds VI. The influence of residue type on papain-catalyzed reactions of some benzoylamino acids with some amino acid anilides. J. Am. Chem. Soc. 75:5539–5542.

Fox, S. W., and Yuyama, S. 1963a. Effects of the Gram stain on microspheres from thermal polyamino acids. J. Bacteriol. 85:279–283.

Fox, S. W., and Yuyama, S. 1963b. Abiotic formation of primitive protein and formed microparticles. Ann. N.Y. Acad. Sci. 108:487–494.

Francis, S., Margulis, L., and Barghoorn, E. S. 1978. On the experimental silicification of microorganisms II. On the time of appearance of eukaryotic organisms in the fossil record. Precambrian Res. 6:65–100.

Fuller, E. C. 1974. Chemistry and Man's Environment. Houghton Mifflin Co., Boston.

Gatlin, L. L. 1972. Information Theory and the Living System. Columbia University Press, New York. p. 1.

Glasstone, S. 1968. The Book of Mars. NASA, Washington, D.C. pp. 193–194.

Granick, S. 1965. Evolution of heme and chlorophyll. In V. Bryson and H. J. Vogel (eds.), Evolving Genes and Proteins, pp. 67–88. Academic Press, New York.

Groth, W., and Suess, H. 1938. Bemerkungen zur Photochemie der Erdatmosphare. Naturwissenschaften 26:77.

Hanson, E. D., 1966. Evolution of the cell from primordial living systems. Q. Rev. Biol. 41:1–12.

Harada, K., and Fox, S. W. 1958. The thermal condensation of glutamic acid and glycine to linear peptides. J. Am. Chem. Soc. 80:2694–2697.

Harada, K., and Fox, S. W. 1965. Thermal polycondensation of free amino acids with polyphosphoric acid. In S. W. Fox, (ed.), The Origins of Prebiological Systems and of their Molecular Matrices, pp. 289–297. Academic Press, New York.

Hardebeck, H. G., Krampitz, G., and Wulf, L. 1968. Decarboxylation of pyruvic acid in aqueous solution by thermal proteinoids. Arch. Biochem. Biophys. 123:72–81.

Hargreaves, W. R., Mulvihill, S. J., and Deamer, D. W. 1977. Synthesis of phospholipids and membranes in prebiotic conditions. Nature 266:78–80.

Hartl, D. L. 1977. Our Uncertain Heritage, Genetics and Human Diversity, pp. 448–450. J. B. Lippincott Co., Philadelphia.

Hartman, H. 1975. Speculations on the origin and evolution of metabolism. J. Mol. Evol. 4:359–370.

Hays, W. L. 1965. Statistics. Holt, Rinehart and Winston, New York.

Horowitz, N. H. 1945. On the evolution of biochemical syntheses. Proc. Natl. Acad. Sci. USA 31:153–157.

Hsu, L. L., Brooke, S., and Fox, S. W. 1971. Conjugation of proteinoid microspheres: A model of primordial communication. BioSystems (Curr. Mod. Biol.) 4:12–25.

Hsu, L. L., and Fox, S. W. 1976. Interactions between diverse proteinoids and microspheres in simulation of primordial evolution. BioSystems 8:89–101.

Hurst, T. L., and Fox, S. W. 1956. The course of proteolysis of lysozyme. Arch. Biochem. Biophys. 63:352–367.

Iben, I., Jr. 1973. Stellar life cycles and cosmic abundances. In M. A. Gordon and L. E. Snyder (eds.), Molecules in the Galactic Environment, pp. 3–20(19). Wiley Interscience, New York.

Jukes, T. H. 1966. Molecules and Evolution. Columbia University Press, New York. pp. 186–187.

Jungck, J. R. 1973. Oligomerization of mononucleotides in anhydrous, hydrous, and hypohydrous milieus: An experimental quest for analogs of protobiogenesis of nucleic acids. Doctoral dissertation, University of Miami, Coral Gables, Fla.

Jungck, J. R., and Fox, S. W. 1973. Synthesis of oligonucleotides by proteinoid microspheres acting on ATP. Naturwissenschaften 60:425–427.

Kenyon, D. H. 1974. Prefigured ordering and protoselection in the origin of life. In K. Dose, S. W. Fox, G. A. Deborin, and T. E. Pavlovskaya (eds.), The Origin of Life and Evolutionary Biochemistry, pp. 207–220. Plenum Press, New York.

Knight, C. A. 1974. Molecular Virology. McGraw-Hill Book Co., New York.

Korn, R. W., and Korn, E. J. 1971. Contemporary Perspectives of Biology. John Wiley and Sons, New York. pp. 457–458.

Krampitz, G., Baars-Diehl, S., Haas, W., and Nakashima, T. 1968. Aminotransferase activity of thermal polylysine. Experientia 24:140–142.

Krampitz, G., and Fox, S. W. 1969. The condensation of the adenylates of the amino acids common to protein. Proc. Natl. Acad. Sci. USA 62:399–406.

Krampitz, G., Haas, W., and Baars-Diehl, S. 1968. Glutaminsäureoxydoreduktase-activität von polyanhydro-α-aminosaüren (proteinoiden) Naturwissenschaften 55:345–346.

Kvenvolden, K. A. 1972. Review of methods used in lunar organic analysis: Extraction and hydrolysis techniques. Space Life Sci. 3:330–341.

Lacey, J. C., Jr., Stephens, D. S., and Fox, S. W. 1976. Unpublished experiments.

Lederberg, J. 1959. Ergebnisse und probleme der genetik, Angew. Chem. 71:473–480.

Lehninger, A. L. 1975. Biochemistry. 2nd ed. Worth, New York.

Lewis, G. N., and Randall, M. 1961. (Revised by K. S. Pitzer and L. Brewer.). Thermodynamics. McGraw-Hill Book Co., New York.

Lipmann, F. 1965. Projecting backward from the present stage of evolution of biosynthesis. In S. W. Fox (ed.), The Origins of Prebiological Systems and of their Molecular Matrices, pp. 259–273. Academic Press, New York.

Lipmann, F. 1972. A mechanism for polypeptide synthesis on a protein template. In D. L. Rohlfing and A. I. Oparin (eds.), Molecular Evolution: Prebiological and Biological, pp. 261–269. Plenum Press, New York.

Mader, S. S. 1976. Inquiry into Life. Wm. C. Brown Co., Dubuque, Iowa.

Margulis, L. 1971. The origin of plant and animal cells. Am. Sci. 59:230–235.

Mark, H., and Whitby, G. S. 1940. Collected Papers of Wallace Hume Carothers on High Polymeric Substances. Wiley Interscience, New York.

Melius, P. 1977. Composition and structure of thermal condensation of polymers of amino acids. In E. E. van Tamelen (ed.), Bioorganic Chemistry, Vol. III, pp. 123–136. Academic Press, New York.

Melius, P., and Sheng, J. Y.-P. 1975. Thermal condensation of a mixture of six amino acids. Bioorg. Chem. 4:385–391.

Mikelsaar, H. N. 1975. A concept of amino acid archaeorelation: Origin of life and the genetic code. J. Theor. Biol. 50:203–212.

Miller, S. L. 1955. Production of some organic compounds under possible primitive earth conditions. J. Am. Chem. Soc. 77:2351–2361.

Nakashima, T., and Fox, S. W. 1972. Selective condensation of aminoacyl adenylates by nucleoproteinoid microparticles. Proc. Natl. Acad. Sci. USA 69:106–108.

Nakashima, T., Lacey, J. C., Jr., Jungck, J., and Fox, S. W. 1970. Effects of several factors on chemical condensation of mixed amino acid adenylates. Naturwissenschaften 57:67–68.

Nakashima, T., Jungck, J. R., Fox, S. W., Lederer, E., and Das, B. C. 1977. A test for randomness in peptides isolated from a thermal polyamino acid. Intl. J. Quantum Chem. QBS4:65–72.

Noguchi, J., and Saito, T. 1962. Studies on the catalytic activity of synthetic polyamino acids having an imidazole group in the active site. In M. A. Stahmann (ed.), Polyamino Acids, Polypeptides and Proteins, pp. 313–327. University of Wisconsin Press, Madison.

Oparin, A. I. 1924. Proiskhozhdenie Zhizny. Izd. Moskovshi Ravochi, Moscow.

Oparin, A. I. 1957. The Origin of Life on the Earth, 3rd ed., trans. by Ann Synge. Academic Press, New York.

Oparin, A. I. 1968. Genesis and Evolutionary Development of Life. Academic Press, New York.

Orgel, L. 1968. Evolution of the genetic apparatus. J. Mol. Biol. 38:381–393.

Oshima, T. 1968. The catalytic hydrolysis of phosphate ester bonds by thermal polymers of amino acids. Arch. Biochem. Biophys. 126:478–485.

Pirie, N. W. 1954. On making and recognizing life. New Biol. 16:41–53.

Ponnamperuma, C. 1972. Lunar organic analysis: Implications for chemical evolution. Space Life Sci. 3:493–496.

Price, C. C. 1974. Synthesis of Life. Dowden, Hutchinson, and Ross, Stroudsburg, Pa.

Racker, E., and Stoeckenius, W. 1974. Reconstitution of purple membrane vesicles catalyzing light-driven proton uptake and adenosine triphosphate formation. J. Biol. Chem. 249:662–663.

Rhodes, W. G., Flurkey, W. H., and Shipley, R. M. 1975. Thermal proteinoids, a project in molecular evolution for the undergraduate biochemistry laboratory. J. Chem. Educ. 52:197–198.

Rohlfing, D. L. 1967. The catalytic decarboxylation of oxaloacetic acid by thermally prepared poly-α-amino acids. Arch. Biochem. Biophys. 118:468–474.

Rohlfing, D. L. 1976. Thermal polyamino acids; synthesis at less than 100° C. Science 193:68–70.

Rohlfing, D. L., and Fox, S. W. 1967. The catalytic activity of thermal polyanhydro-α-amino acids for the hydrolysis of p-nitrophenyl acetate. Arch. Biochem. Biophys. 118:122–126.

Rohlfing, D. L., and McAlhaney, W. W. 1976. The thermal polymerization of amino acids in the presence of sand. BioSystems 8:139–145.

Ross, H. H. 1962. A Synthesis of Evolutionary Theory. Prentice-Hall, Englewood Cliffs, N.J. pp. 45–48.

Ryan, J., and Fox, S. W. 1973. Activation of glycine by ATP, a divalent cation, and proteinoid microspheres. BioSystems 5:115–118.

Samuel, E. 1972. Order: In life. Prentice-Hall, Englewood Cliffs, N.J.

Schmitt, F. O. 1956. Macromolecular interaction patterns in biological systems. Proc. Am. Philos. Soc. 100:476–486.

Schopf, J. W. 1976. Are the oldest 'fossils' fossils? Orig. Life 7:19–36.

Schwartz, A. W., Bradley, E., and Fox, S. W. 1965. Thermal condensation of cytidylic acid in the presence of polyphosphoric acid. In S. W. Fox (ed.), The Origins of Prebiological Systems and of their Molecular Matrices, pp. 317–325. Academic Press, New York.

Snyder, W. D., and Fox, S. W. 1975. A model for the origin of stable protocells in a primitive alkaline ocean. BioSystems 7:222–229.

Stillwell, W. 1976. Facilitated diffusion of amino acids across biomolecular lipid membranes as a model for selective accumulation of amino acids in a primordial protocell. BioSystems 8:111–118.

Szent-Györgyi, A. 1977. The living state and cancer. Proc. Natl. Acad. Sci. USA 74:2844–2847.

Tien, H. T. 1974. Bilayer Lipid Membranes. Marcel Dekker, New York. pp. 457–460.

Turcotte, D. L., Nordmann, J. C., and Cisne, J. L. 1974. Evolution of the Moon's orbit and the origin of life. Nature 251:124–125.

Usdin, V. R., Mitz, M. A., and Killos, P. J. 1967. Inhibition and reactivation of the catalytic activity of a thermal α-amino acid copolymer. Arch. Biochem. Biophys. 122:258–261.

Vallery-Radot, P. 1922. Oeuvres de Pasteur. Tome II: Fermentations et Generation dites Spontanees. Masson et Cie, Paris.

Vegotsky, A. 1972. The place of the origin of life in the undergraduate curriculum. In D. L. Rohlfing and A. I. Oparin (eds.), Molecular Evolution: Prebiological and Biological, pp. 449–458. Plenum Press, New York.

Volpe, E. P. 1977. Understanding Evolution. 3rd Ed. Wm. C. Brown Co., Dubuque, Iowa.

Waehneldt, T. V., and Fox, S. W. 1967. Phosphorylation of nucleosides with polyphosphoric acid. Biochim. Biophys. Acta 134:1–8.

Waehneldt, T. V., and Fox, S. W. 1968. The binding of basic proteinoids with organismic or thermally synthesized polynucleotides. Biochim. Biophys. Acta 160:239–245.

Wald, G. 1954. The origin of life. Sci. Am. 191(2):44–53.

Weber, A. L., Caroon, J. M., Warden, J. T., Lemmon, R. M., and Calvin, M. 1977. Simultaneous peptide and oligonucleotide formation in mixtures of amino acid, nucleoside triphosphate, imidazole, and magnesium ion. BioSystems 8:277–286.

Weinberg, S. L. 1977. Biology, An Inquiry into the Nature of Life. 4th ed. Allyn and Bacon, Boston.

Winchester, A. M. 1976. Heredity, Evolution, and Humankind, pp. 58–69. West Publ. Co., St. Paul.

Winchester, A. M. 1977. Genetics. 5th ed. Houghton Mifflin Co., Boston. pp. 410–411.

Wolfe, S. L. 1977. Biology: The Foundations. Wadsworth Publ. Co., Belmont, Cal. pp. 366–368.

Wolken, J. J. 1975. Photoprocesses, Photoreceptors and Evolution. Academic Press, New York.

Wood, A., and Hardebeck, H. G. 1972. Light enhanced decarboxylation by proteinoids. In D. L. Rohlfing and I. A. Oparin (eds.), Molecular Evolution: Prebiological and Biological, pp. 233–245. Plenum Press, New York.

Yoshida, M., Sone, N., Hirata, H., Kagawa, Y., Takeuchi, Y., and Ohno, K. 1975. ATP synthesis catalyzed by purified DCCD-sensitive ATPase incorporated into reconstituted purple membrane vesicles. Biochem. Biophys. Res. Commun. 67:1295–1300.

Young, R. S. 1966. Extraterrestrial Biology. Holt, Rinehart and Winston, New York.

Yuki, A., and Fox, S. W. 1969. Selective formation of particles by binding of pyrimidine polyribonucleotides or purine polyribonucleotides with lysine-rich or arginine-rich proteinoids. Biochem. Biophys. Res. Commun. 36:657–663.

Zuckerkandl, E., and Pauling, L. 1965. Evolutionary divergence and convergence in proteins. In V. Bryson and H. J. Vogel (eds.), Evolving Genes and Proteins, pp. 97–166 (162–3). Academic Press, New York.

chapter 3
The Web of Life

Bernard M. Loomer

Graduate Theological Union
Berkeley, California

This chapter addresses an ancient theme, a theme expressive of an ancient wisdom and symbolic of an ancient vision. The theme is the unity, the interconnectedness, the interdependence, of human life, even the unity of all life. This topic has had several historical origins. It has been baptized by a plurality of faiths. It has been called by many names. I call it the web of life. In the Judaic-Christian tradition of the west, dimensions of this theme have emerged through an evolved understanding of covenantal relationships and the development of the concept of the kingdom of God. This unity was described as at once a present realization and a possibility. The kingdom is both here and yet coming.

On the one hand, ancient wisdom and revelatory disclosure proclaimed the unity of life to be a primordial condition of our existence, a stubborn fact of our historical experience. It is present as a given structure of our common life. As a network of dynamic relations, this web is also creative of that common life. This insight was asserted and maintained in the midst of the seemingly almost overwhelming evidence against it. To overstate the matter perhaps, this insight emerged in spite of and because of the presence of persistent and destructive conditions that seemed to negate its possible truth.

But this insight did not emerge easily. Nor was it maintained blindly or simply hopefully. The profoundest minds and the most deeply tested spirits among our early forebears knew from prolonged and sometimes tortuous inquiry that at the level of ultimate questions truth is hard won. It has its price, and sometimes the cost is very high. These forefathers of ours were deeply empirical. They knew that evidence is a function of perception. But they also knew that in basic matters perception in turn is a function of disciplined sensitivity and discernment. They knew, furthermore, that this kind of discernment is a product of profound searching involving patience, stamina, courage, and single-mindedness, which are qualities of a sustained trust. They were keenly aware that intuitions of those values from which the deepest meanings are derived are not granted to the successful, the rich, the powerful, the simple minded, the dilettantes, the fainthearted, or the untested. They learned long ago a methodological principle, which Whitehead has reiterated in a more abstract fashion, namely, that with respect to the most important and most difficult issues persuasive evidence is neither derived nor derivable from simple and unambiguous data that are clearly and distinctly given. In the language and substance of a Biblical parable, the wheat and the tares grow together. On the deepest issues, evidence is often accessible only to those who can discern the wheat that exists not apart from but in the intimate association with destructive tares and empty husks.

On the other hand, ancient imagination and prophetic vision proclaimed the unity of life to be a state of affairs that was in the making. All life, at least at the human level, is interconnected. That was the perceived fact. But life was filled with evil, injustice, corruption, poverty, disease, and misery of all sorts. Thus, the full flowering of the abundant life that was possible within this web of interconnectedness lay in the future, a kingdom of goodness and fulfillment that was coming. The questions as to when this kingdom was to arrive and by whose action (God's or man's) are not for the moment decisive. But it can be noted that the basic pattern of life that was projected as illustrative of existence within the kingdom that was coming was continuous with the underlying and primordial web of creative relatedness that was discerned as already present in the midst of the goodness, joy, evil, and agony of the world that then existed.

The unity of life is not just an ancient theme and dream. It is a perennial topic, and it is grounded on a recurring insight. This insight asserts our common and universal interdependence. But this insight has had an uncertain, vague, and at times a hidden career. It has been preserved in our several religious traditions, even though for much of their history it was lost to the consciousness and conscience of the institutions that carried these traditions. The personal and institutional representatives of these traditions, down to and including the present day, have too often been the most flagrant violators of this insight and vision. It has been kept alive in the works of some of the most sensitive and brooding artists of words, paint, and music, particularly the poets, even or perhaps especially when their rhymed portrayals of our faithlessness and inhumanity toward each other have been most biting and ironic. "Never ask for whom the bell tolls. It tolls for thee."

This insight and the possibility or hope of its practical realization has been the concern of some philosophers. In at least its political, economic, and social aspects it seems to constitute part of the driving force of Marxist thought. It might be said, but perhaps only in the presence of some qualifications and disclaimers, that this insight concerning the interconnectedness of human life lies hidden within the value assumptions by which democratic political theory understands and justifies itself. I would contend that the American experience can be interpreted as a slow, uncertain, ambiguous, and resistance-ladened evolution toward the acknowledgment and embodiment of this insight. In the profoundest moments of its self-understanding America knows that this insight is the heart of the American dream.

But on the whole, the many theologies and philosophies (religious, metaphysical, political, and economic) that have intervened between ourselves and the ancient world have forgotten, distorted, or denied this basic insight. These theologies and philosophies have not adequately reflected the central importance of the web of interdependence. In every aspect of our lives we are poorer for this failure.

The neglect of this insight in the premodern ages, in both their theoretical understandings and their institutions, was not as serious in its total consequences as ours has been. The destructive as well as the creative resources available to these earlier ages lacked the magnitude of power that is within our possession.

I

The scientific and technological revolutions have changed the faces of both nature and culture. The consequent economic and industrial developments, together with the problems of population, food, energy, the magnitude of military expenditures, and the limitations

of natural resources, have forced us to at least a preliminary awareness of some aspects of our interdependence. Modern methods of
transportation and communication have shrunk the spatial and
temporal dimensions of our planet. They have also brought the
peoples of the world face to face, and from these encounters the
vision of one human family has emerged as needed, eminently
desirable, and theoretically possible. The revolution of political,
economic, and social expectations of all peoples, in the face of our
technological problems and limited resources, makes the commit-
ment to the realization of this vision mandatory.

Thus, through the pressures of modern circumstances, we
have come to recognize, albeit grudgingly, the relevance of an
ancient dream, however distant its feasibility may be.

Now this acknowledgment of our interrelatedness, however
belated its emergence in our self-understanding, is surely preferable
to the absence of this acknowledgment. But this awareness, which
has been forced on us in spite of ourselves, may be not only belated but also merely skin deep in its grasp of what is involved.
The recognition of our interdependence, where this is understood
simply as a modern phenomenon, means that we have not perceived
and absorbed the insight that undergrids this ancient vision whose
realization has now become a modern necessity. Our failure to
appreciate the fact of our interdependence in some elemental depth
may contribute to our making repeatedly inadequate responses to
the baffling problems that threaten to engulf us.

We are not interdependent simply because of the technological
circumstances of the modern world. In the elemental sense, we
have not suddenly become interdependent whereas formerly we
were essentially independent, both as individuals and as societies.
We have been and we are now interdependent because this interrelatedness is an ever-present and inescapable structure within
life itself. We have become more aware of our interdependence
because of contemporary social conditions, but these conditions
have not created the basic fact of interrelatedness that is founda-
tional for all human life.

More than any previous age, the modern world has disclosed
depths and dimensions of our interdependence. These modern
social and technological conditions have concretized the fact of our
interrelatedness in ways and to a degree far exceeding those of the
premodern world. The fact of our interdependence has been more
fully actualized and exemplified. Our lives have in truth become

more concretely interwoven. In this sense we have become more interdependent.

These modern conditions have created specific types of relationships, which exemplify certain styles of interdependence. They have given birth to concrete social relationships that illustrate particular forms of the web of interconnectedness within which we all live. The particular form of the web, or the particular pattern of interrelatedness, may be understood as an emergent from these conditions. But the fact of the web, or the fact of our being interdependent, is not itself an emergent. It is a condition indigenous to life itself.

These newly disclosed depths and dimensions of our interdependence contain possibilities for greater creative growth or for more devastating destruction. Because of our more concrete interdependence, and because of our greater power for good or ill, we are required to be more ethically sensitive in the larger as well as in the more personal relationships of our existence than has been demanded of previous generations. The fulfillment of this requirement involves the creation of individuals and societies with a stature of mind and spirit adequate to the monumental challenges that confront us. The movement toward this goal must become, I suggest, at least one of the basic aims of our educational and nurturing institutions.

It appears to be the case that with respect to the important affairs in our lives most of us learn the hard way. It seems that most of us are willing to learn only when we are forced to, only when failure or disaster or tragedy has occurred, only when there seems to be no alternative if life is to yield any appreciative degree of satisfaction. Perhaps this is the way things are much or most of the time, even for the best of us. This may be one of the ways, or perhaps the only way, that wisdom is produced. But too often this method results in there being too many instances of "too little and too late."

Probably most of us cry only when we ourselves are hurt and suffer, just as most of us seldom if ever give up our power voluntarily. We reduce our claims only when someone else is strong enough to force us to do so. But if we cry only when we ourselves are hurt, if we suffer only in our own misfortune or agony, if we are motivated to respond only to our own need, if our sensitivity and compassion reach out only to the boundaries of our enclosed selves, if the plight of another is not in itself a wholly adequate reason for us to extend our hand and spirit to the other, if we do not see our own shrinkage in another's humilitation—then we have missed the whole point of the Gospel's parables about the fact and nature of the web.

As a species we seem to be somewhat better at handling crises once they occur than in preventing their occurrence. But with respect to some of our problems it may be the case that the crisis consists in preventing the occurrence of those conditions whose destructiveness may become unredeemable.

II

During the period of the growth of our technological interdependence, certain intellectual advances have occurred that confirm and also extend this ancient insight. These intellectual developments may help us to understand and to grapple more creatively with the technological problems that in part define our contemporary world.

The rise and evolution of the social sciences is a phenomenon of the twentieth century. From these disciplines we have derived an understanding of the societal conditioning of all aspects of our lives. The social is not something added to what has been regarded as individual. Nor is the social to be interpreted as the amorphous aggregate of the individual members of a group. The social is a complex, historically created formative structure in which all the members of a society participate and from which they derive a partial and yet common identity. These disciplines have given us social conceptions of various aspects and types of reality, extending even to the notion of social conceptions of reality as a whole. The existence of the discipline of social psychology, as represented, for example, in the work of George Herbert Mead, is symbolic of the change in the intellectual climate that is characteristic of our age. (The work of Freud is important with respect to other features of twentieth century life, but it is not decisive for understanding the social or relational character of our existence.)

Part of the groundwork for this development had been laid in the nineteenth century with the rise and evolution of that mode of mind that is called historical understanding. One stage of that development involved the notion that any historical phenomenon must be viewed contextually. This methodological principle did not necessarily imply (although this implication was drawn by some historians) that a historical individual could be totally accounted for by a knowledge of his environment. But it did mean that the historical individual was what he or she was only in, and in large part because of, that particular context.

Meanwhile there arose a line of philosophical thought that began, roughly, with William James and culminated in the work of Whitehead. In contrast to the idealists and the early empiricists with whom he struggled, James insisted (among other matters) that the connections between things were given and known to us in experience, along with the things connected. The connections were not in the first instance furnished by the knowing mind. Whitehead's early involvement in relativity physics led him eventually to a philosophical outlook that emphasized (with respect to the purposes of this chapter) two fundamental principles: 1) the ultimacy of the process of becoming, and 2) the primacy of the relatedness of things.

The principle of the relatedness of things means that there are no self-existing, completely independent things in our world. Things exist only in contexts or fields of energy or environments. They exist in contexts in the profoundest and ultimate sense. They are literally largely derived from these fields of energy.

The principle of becoming refers to a process whereby individuals are created, or emerge as individuals, by synthesizing causal influences vectorally derived or projected from these fields. In this process of synthesis these emerging individuals contribute to their own creation. This is the independence and solitude of their subjective life. This is the freedom of their self-creativity whereby they achieve whatever fulfillment they can realize. This fulfillment is the measure of their contribution to the society that gave them birth.

Out of this confluence of developments has come the conception of the communal individual or the social self, or what I would call the relational self. This view of the self is a pivotal contribution of the twentieth century to the study of human nature. But, while this concept of the social self is by now something of a commonplace in the academic world, I think that it is not understood in the most basic and radical sense.

The concept of the social self does not mean simply that the self lives in a particular society upon which the self is dependent in some external fashion. The social self is not an independently existing individual who has social relations with other members of society because the self requires the presence of other selves in order to realize its possibilities. The concept means not only that the self has its existence solely within a society but also that the society lives within the self. In saying that an individual is shaped by his or her society, we mean that this is to be understood in something more than a nominal sense. Whatever you are influenced by, whatever shapes or determines you, becomes internalized. It lives within you. It exists as a formative energy within you. It is quite literally a part of you. The concept of the social self, at least as I conceive of it, refers to a self whose very being, whose inner life, is largely derived from and constituted by its relations with its society. From this perspective, the self does not *have* experiences as though the self transcended its experiences. Rather the self *is* its experiences. Analogously, the social self does not only *have* relationships with others. By and large the self *is* its relationships. Or, rather, the individuality of the self consists in what it does with the causal influences it receives from others. In this sense the self is an emergent from its relationships. In Mead's terms, the self cannot become an individualized "I" until it has first become a socialized "me."

This process of the self's emergence from its relationships extends over the entire historical life of an enduring individual. In the course of its life the emergent self exercises its causal influences on other selves. In this fashion a society is derived from the mutual constitutive relations that exist among its members.

The point can be expressed in slightly different terms. The principle of the primacy of relations means that these constitutive relations are the causal influences from which we originate and from which our existence as a self begins. These relationships are the carriers of the qualitative energies that are experienced as the events and occasions of our concrete daily existence. In these terms, it is not the case that we are fully formed and independently existing individuals who entertain various kinds of relationships with other similarly constituted individuals. Rather, we live and actualize our individuality only within relationships. The self exists as a subjective center or a unified focus of dynamic relations within an extensive field. We who live in the traditions of the west tend to think that the individual is the basic unit of reality. In certain respects this appears to be the case, especially since, at least at the ordinary human level, the individual seems to be a clearly demarcated, encompassable, and separable organism that can be rather sharply distinguished from its environment. But in other respects the concrete individual is an abstraction from this extensive and interconnected field of activity. In these latter respects, to overstate the matter perhaps, the social entity of the extensive and interconnected field seems to be the unit of reality.

This is an overstatement in terms of several considerations, not the least of which is the point that the spatial and temporal extensiveness of a field may not be determinable in any humanly relevant way. This way of putting the point raises a host of questions that I am not prepared to deal with in any adequate fashion. However, I am concerned to sustain an emphasis. On the one side, the field exists only in terms of its constituent members who emerge as individualized centers of unified activity. On the other side, the influences issuing from any individual permeate the field in some indefinite fashion, and in this sense an individual exists throughout the interconnected field. From this perspective, an individual considered apart from his or her relational context, or a contextual field viewed without reference to its members, is an abstraction from the concrete togetherness of actuality.

This understanding of the matter does not mean that the individual self is swallowed up by its relationships. The self is constituted largely by its relationships, but it is not wholly a function of them. The unique individuality of the self derives from what it created out of what it has been given. Yet, even here, the solitude of the self is not a dimension of life that is encountered apart from its relationships. It is experienced most poignantly in the midst of contextual relationships. It emerges as a kind of transcendence of relationships that yet has its being only within them. The awesomeness of solitariness arises from the awareness that the individual self, however close its relations with others, is not these others but is finally its own singular reality. The more intimate the relationship, the deeper the sense of solitariness.

III

If the basic point of these remarks is imaginatively or philosophically, even if not scientifically, generalized, we arrive at the notion that the world may be viewed as an indefinitely extended field of interconnected events. To my mind this is the single most comprehensive and significant statement to be derived from the general point of view within which this chapter is written.

At many junctures these connections may seem to be, and may in fact be, vague, obscure, tenuous, or even absent. In some important respects, this field may be quite loosely organized and can be called a field only gratuitously. But this general point of view sees the world as an open-ended, evolving, unfinished, and deeply problematic world, what Loren Eiseley refers to as an "unexpected universe." This is a relativized world of deep ambiguities and ragged edges, where we operate perhaps only within "margins of intelligibility," to borrow Bernard Meland's descriptive terms. This may be a world in which actuality itself apes the history of science, wherein new dimensions of life are disclosed, and suggested answers lead on to more complex problems. In such a world we can live without the false resolutions provided by "final" answers, provided the questions enhance the zest for life.

But, granted the tentativity and the unease that accompany the making of this kind of generalization, I suggest that the generalized conception of an interconnected and socialized field is a modern and naturalistic version of the web of life—granted the abstractness and the obvious incompleteness of this description.

IV

The general topic of the XIIIth Nobel Conference, "The Nature of Life," would seem to require a more concrete treatment of some aspects of human life within this web. That phase of the topic is considered in the following section.

I preface this discussion with a general statement concerning the subject of order. The guiding principle here is that you do not explain the concrete by the abstract. Rather, you explain the abstract by the concrete. To the extent that there is an order of nature and life, this order does not obtain because events obey some imposed law. Neither is it the case that events are interconnected because there is a common order. Rather, there is an order because events are interrelated. The order, as such, is an abstraction from the relationality of events. The order reflects the relationality of the web, not the other way round. To express the point differently: we are not interrelated because there is a law or principle of love. There is a principle of love because we are interrelated.

The givenness of any order with respect to any society or specific field of activity testifies to a relatively stable and recurring pattern of interconnection between the events that constitute that society. To this extent at least, order is an emergent from concrete relationships.[1]

This principle can be expanded. If reason is correlative with order or structure, then reason has its source in the interconnectedness of things, not the reverse. Whatever reasonableness life exemplifies is derived finally from the concrete and strange workings of the web. The degree to which life makes sense cannot be determined in the last analysis by the degree to which the operations of the web conform to some preestablished criteria of meaning. The question of the meaningfulness or absurdity of life is referent to the web itself. There is no higher court of appeal.

The fact of the web means that we are members one of another. In this relational mode of life we belong and participate in the web because we exist. Through our activities we help create the web. The web in turn gives birth to us. The influences flowing from any point within this field reverberate throughout the web in varying degrees of intensity. The increase in relational value for one individual is the enrichment of all. The lessening of one is the diminution of all. The faithlessness of one adds to the improverishment of each. As individuals we are fulfilled through, with, and in others.

Now one may concede the existence of the web and yet come to a thoroughly sceptical conclusion. For some, perhaps for many people, the evidence for the existence of the web seems to be derived from dehumanizing and life-denying experiences. It must be granted that the fundamental character of the web, namely the interdependence of its members, is illustrated just as surely by the mutual destructiveness that is so much a quality of our common life as it is exemplified in our more constructive efforts to actualize the human family and to enhance the quality of its life. Increasingly we seem to be creating a sense of nothingness in and for each other. One may acknowledge that the process of evolution from simpler to more complex forms of life may be said to embody, in general, a creative direction and thrust, and yet wonder whether

---

[1]The question as to whether there is a most general, ultimate, primordial, and non-emergent order, which all orders of less generality illustrate, and which all events do and must exemplify (in varying degrees), is too complex a topic to be discussed within the limits of this chapter. Suffice it to say that many past conceptions of this kind of order have not involved the primacy of the interconnectedness of events.

this process has reached another dead end with the appearance of the human species.

I believe that this point must be taken with great seriousness. Yet, because of and in spite of this consideration, and in contrast to those who believe that this evolutionary process operates pretty much in terms of blind and fortuitous factors, I hold that there is a creative aim within the web, including the present life of man. This aim is inherent within the web although it is manifested only in the midst of uncreative, ambiguous, and destructive conditions and forces.

How is this creative push to be seen? It can be seen most fully in the living of the relational life in its depths. The resources for the living of this kind of life are experienced only in the living of it. The exemplification of this type of life calls for and also creates people, societies (and institutions) of greater stature. These are people large enough to absorb the violence, hatred, contradictions, and emptiness of modern life without being overcome. The beatitudes of the New Testament are descriptions of the fulfillment and resources available to those who attempt to live the relational life in the midst of extreme contradictions that border on chaos and nothingness. This constitutes at least the beginning of an answer to the despair and nihilism of our age.

Yet the powerful undercurrents of the flow of existence remain to sober us. I have said that the aim of the web is the creation of individuals and societies of greater stature. But the odds seem to be weighted heavily, almost overwhelmingly, against the realization of this aim. We cannot avoid wondering whether the creative resources are adequate for the task. Some of the problematic features of our lives can be noted. The following considerations do not include an analysis of the large social structures that dominate our daily lives.

There is, in the first place, something deep within an individual that militates against the enrichment of the web and the achievement of his finest fulfillment. The individual is born in the web, he is largely composed of the relationships derived from the web, he is sustained by it, and he is basically so constituted that it can be said that he is built for the relational life. Yet, having emerged as an individual from this web of interdependence, he is prone to feel that the actualization of his individual existence resulted in a stoppage in the causal chain of interdependence, as though he were self-existent, and as though his very being were his possession exclusively.

The complicating fact is that he is independent as well as dependent. He is free as well as determined. As an organism that is a distinguishable and centered unit with its own life and sense of value, he becomes almost totally absorbed with his own feelings and needs and sense of worth. He becomes primarily concerned with the realization of his possibilities. He becomes self-obsessed. He makes himself the single-track focus of his existence. Others for the most part exist in their humanity only to the degree that they serve his own ends. He comes to believe that above all he must look out for himself, because no one else will—not really. He thinks fulfillment means that his interests must be met on his terms, because in his mind no one else is at heart concerned with his unique subjective hopes and feelings. He thinks he can trust only those relationships that are grounded in himself as a center. He finds it difficult to give himself to those relationships that involve a mutual giving and a mutual receiving—relationships that can increase the stature of all those so related but that cannot be controlled. In his self-centeredness he impoverishes himself and the web.

Second, pride in one's self is an important ingredient in the equipment of a well-functioning person. It is basic to his sense of self-worth. Without this self-affirmation, at least to a reasonable and ordinate degree, he is unable to participate in creative relationships. Without this basic affirmation of himself, his life-stance may become that of self-hatred, and this in turn can lead to a deep apathy or a destructive violence within his spirit.

But ordinate and justifiable pride is an unstable quality. It is inflated quite easily and then is transformed into something inordinate and destructive. The person whose sense of self-worth is interlaced with the graciousness of humility knows that his self-worth involves the labors and gifts of others as well as those of his own. He gratefully acknowledges his dependence on others without feeling demeaned. But many of us, in our insecurity, believe that a sense of self-worth that is in part derived from a relation of dependence or interdependence is not an adequate measure of our true worth. We want to be acknowledged for what we are in and by ourselves. We want to feel that we deserve our status because of our own merits. We want to be recognized for what we have truly earned.

In this spirit we promulgate the belief that life is or should be a matter of elemental justice, that we should be rewarded for what we ourselves have truly earned, and that those who have not paid their proper dues should not belong to the club. This credo of the self-made and self-deserving individual involves the corollary that those who are filled with the bounties of life deserve them because they have paid the price. The implication seems to be that those who are not so favored deserve only the little they have, because of their sinfulness, or their laziness, or their primitive state. To them the hand of charity may be extended.

In this stance of the human spirit (which involves an inadequate sense of the role of luck in human affairs and displays an undeveloped sense of irony) we become defensive, self-justifying, and unresponsive. We are able to accept and befriend only those who have earned their station in life and who are of a like mind with our own. Because in our pretentious pride we do not recognize our need to repent or to be forgiven, we are unable to extend our forgiveness to others. In these ways we estrange ourselves from the web and from ourselves. Our pride, which at its base is symbolic of our strength in the web of our interrelationships, has become demonic. It shreds the web.

My focus here has been on individuals. However, these remarks apply also to nations and smaller societies. At the communal and national level this need for a sense of self-worth can lead and has led to the notion of a specially favored people, and favored on their own self-justifying terms.

This understanding of life, which is grounded on a nonrelational view of pride and a nonsocial conception of the self (as well as a nonsocial interpretation of reality in its larger dimensions), is an ancient and perennial theology. The wrestling with this perspective absorbed much of the attention of the ancient Hebrews, as the book of Job, for example, illustrates. This viewpoint persists as a powerful element in our American self-understanding. It corresponds to something deep within the human soul. The evolution of the spirit to a more adequate view is painful and slow.

Our religious tradition treated this interpretation of life with impressive seriousness. Our forefathers regarded the basic attitude exemplified in this life-stance as one of the strongest of those conditions that are inimical to the reception of the good news of the kingdom. They emphasized the role of the grace of God in human salvation because it was felt that an individual, in his effort to save himself, would develop a pride that would estrange him from both God and his fellows.

The story of the prodigal son is not merely one parable among others. In one sense it is *the* parable of the New Testament. The elder brother speaks for all of us who believe that one must earn his way into the web of the human family. The father's celebrative reception of his repentant and returning son is an affront to those of us who have been faithful to our obligations over long periods of time. The notion that one can be readmitted to the web by an act of repentance, or that from the point of view of the web the son had never lost his membership or his rights, seems to make a mockery of virtue. We who have earned our place and paid our dues now feel estranged. Our years of loyal service have ended in bitterness, and all that we thought we had is as nothing.

The parable of the prodigal son reappears as the parable of the owner of a vineyard who paid men who worked for one hour the same wages he had contracted to pay to those who worked all day. This is not a blueprint for an economic system, but a parable of the web. Participation in the life of the web is at the behest and generosity of the web. It is not to be determined on the basis of justice. Why should those who have worked longer resent the admission to the web of those who have worked less? The generosity of the web is such that we are all members of it. All of us are admitted on the same terms. I suggest that this is one of the hardest parables with which to live.

V

In the third place, it is not just our individuality and our pride that may slash the fabric of the web of life. Our freedom, which is a prerequisite of our humanity, can be misused so as to lead us to dehumanizing consequences.

Freedom has several dimensions and meanings. Most fundamentally, perhaps, it is an impulse to create, to create ourselves by the actualizing of our possibilities, and to move beyond our present situation. In this respect freedom is expansive. It is a self-transcendence. It may also be a transcendence of nature and of other people, and this status may involve the use of them as instruments for the realization of our own ends. And if we work hard enough, granted the presence of adequate resources (both natural and human), or if fortune smiles on us, we may become successful—even wealthy and powerful.

I need not recite the many ways in which the possession of wealth and power, and the achievement of success in terms of any established system of rewards, become agencies of estrangement. The instances that document this fact are beyond measure. This condition defines the profoundest problem that confronts American life. Our long-run response to it will determine the outcome of the American experiment. In response to the question as to how a rich man could enter the kingdom of heaven, the answer as reported in the Gospels was radical: he must give up his wealth and power.

This answer seems to be not only absurd but fantastic in its degree of irrelevance. Yet no economic system as such is self-validating. Its justification is grounded upon considerations about values that transcend the range of economic activity. Elsewhere I have indicated that the traditional conception of unilateral power is not adequate for today's problems. I have suggested a view of power as relational, which is, to my mind at least, more coherent with the relational life of the web. We also need a conception of economic activity that reflects the primacy of this kind of life.

Our virtues are limited by the dimensions of life they encompass. Their adequacy is thereby circumscribed. To this extent they are strengths. But in choosing to justify ourselves in a nonrelational manner, we become defensive. We claim too much on our own behalf. We ignore or deny the limitations of our virtues and overplay our strengths. In this fashion our virtues become vices and our strengths become weaknesses.

It is because of the frequent and perhaps almost inevitable misuse of our freedom in this fashion that the attitude of repentance is one of the primary attributes of life within the web. The Gospels first depict Jesus as announcing a coming kingdom and calling for the peoples' repentance, repentance for their sins and the pretensions of their goodness. As self-righteousness and arrogance are two of the ugliest and deadliest enemies of the web of the human family, so the humility of contrition and the graciousness of forgiveness are the oils that lubricate the operations of the relational life.

VI

Fourth, there is an objective condition within the web itself that is of major significance. The members of my generation owe several debts to Reinhold Niebuhr. One of his most important contributions is his insight that every creative advance brings with it the possibility of greater evil. I take this to be a disclosure of the dialectic of the web of life involving its pervasive interdependence. The insight forces on our attention the sobering realization that there is no progressive conquest of destructive forces. On the contrary, we confront the prospect that every gain requires the presence of greater creative resources to overcome the threat of greater loss.

When this principle is applied to our contemporary world with its technological revolution and the resultant external togetherness of its citizens, our problem of first and enduring priority is that of the quality of life that is available to us. Life may exemplify greater qualitative richness, deeper impoverishment, or more destructiveness. The urgency of the issue has been a pressure of long duration.

I suggest that the adequacy of our response to this problem may be evaluated in terms of our capacity to undergo a rather radical transformation of our understanding, and to shape the personal and social structures of our existence so as to reflect the primacy of the relational network of our interdependence.

This web is the primordial covenantal relationship, a covenant in which all peoples have membership as their birthright, and for whose enrichment all peoples are chosen. Special historical covenants, religious or secular, of less generality are finally justifiable in terms of their contribution to this more inclusive community. The evolution of the human spirit consists in the emergence of a deeper understanding and exemplification of this elemental covenant.

Our history would seem to indicate that this evolution is at best doubtful. Its forward movement is contingent on the creation of individuals and societies with sufficient stature and stamina to sustain the incredible discipline of the relational life. Yet it is only in the context of this style of life that this kind of stature emerges. The life that is required is one lived in a world many of whose forces seem to be arrayed against the realization of the level of life that is needed. Yet these agencies also exist within the web.

We should be aware of what we are asking of ourselves. In this chapter I have not attempted to describe, systematically and in detail, the character and qualities of the relational life. That task would require another paper. However, I suggest that our awareness of the price involved may be heightened by looking at one version of the relational life: the incredible history of the Jewish people.

How is it that, in the course of development, ongoing behavior remains attuned to the incessant changes occurring in those features of the external world that are so vital to survival and reproduction? As an ethologist concerned with trying to extend our understanding of the natural behavior of animals, the question on which I want to focus in this chapter is what are the roles of nature and nurture in the ontogeny of the ability to perceive the external world? By *perception* I mean not just the ability to sense changes in the environment, but also to organize and interpret that which is sensed, and to act on the basis of those interpretations.

The problem has long fascinated philosophers and psychologists as well as biologists, not the least because there are many obvious discrepancies between the picture of the external world that we derive through our senses and the reality we know from science. Optical illusions are an obvious case of differences between the real world and the perceived world that, carefully analyzed, have yielded profound insights into how visual perception works (Ratliff, 1965; Gregory, 1974). The development of the ability to see has always had a particular fascination for thinkers engaged in what is still a major philosophical controversy (Hochberg, 1962).

On the one hand are the nativists, with such classical champions as Descartes, Kant, Spinoza, and even Plato believing that much knowledge about the external world and how to process impressions of it is inborn. On the other hand are the empiricists who view perception as a purely learned ability. By this view there is a gathering of impressions into memory if they prove meaningful, and a discarding of those that prove redundant or irrelevant, so gradually deriving a kind of orderliness from the patterns of the organism's interaction with the world in which it lives. The classic empiricist metaphor is that of a blank slate on which there is a gradually accumulating history of the past experiences of the external world of each individual. This picture of the relevant aspects of past environments provides the frame of reference for our future actions, our thoughts, and our strategies for dealing with new situations that we have yet to encounter. The empiricist case was argued eloquently more than two centuries ago by philosophers such as John Locke (1690) and Bishop Berkeley (1709, 1710). The empiricists face problems, however. Higher organisms are confronted with stimulation from the external world that is incredibly complex and disorderly. In coping with this problem, animals and people sometimes seem to use definite rules for the process of stimulus abstraction and generalization involved in their perceptual development (e.g., Herrnstein, Loveland, and Cable, 1976). As with language development, such widespread or "universal" rules are an immediate invitation to invoke something akin to innate knowledge.

One would think that, with several centuries of thought and a good deal of research on an issue so basic to man's view of himself and his place in nature, the problem would have been resolved long ago. Yet precisely equivalent controversies still rage, with no clear prospect of resolution. For example, in a recent volume on the philosophy of language assembled from the proceedings of a symposium on innate ideas there is an illustration from the field of linguistics. Noam Chomsky and his critics argue about whether studies of linguistic competence, with similar rules occurring in different cultures, and linguistic rules developing in children without any *obvious* opportunities to have learned them require us to postulate innate ideas as a basis for the development of language, knowledge, and thought (Searle, 1971).

An equivalent controversy at a more physiological level is now raging about the early development of mammalian vision. Some investigators stress the remarkable, and apparently inborn, abilities of the newborn to process certain kinds of visual information. Others, working in a somewhat different vein, stress the apparent plasticity of the brain's developing ability to analyze what it sees (Hubel and Wiesel, 1963; Barlow, 1975; Grobstein and Chow, 1976; Blakemore, 1977). There is no clear resolution yet in sight. It is a problem with which it is logically difficult to deal, perhaps in part because of our tendency to think only in terms of extreme alternatives. Either perception is innate or it is learned.

There are those who have argued in a middle way, with innate factors influencing the learning process, and of course at a certain level this is so obvious as to be trivial. Clearly no empiricist, however extreme, visualizes an organism that is potentially capable of perceiving all possible changes in the external world. Limits are set by the structure of its sense organs. The sensory world of an insect is obviously different from that of a mammal. We all know of the honey bees' ability to see the intricate patterns of ultraviolet coloration on many flowers, which are invisible to us without the trick of photographing them through a quartz lens that allows the ultraviolet to pass (von Frisch, 1967). It is impossible for us to imagine the subtleties of ultrasonic echolocation at which bats are so incredibly adept (Griffin, 1974). There is recent evidence that birds may be able to hear subsonic sounds, so low pitched that we cannot hear them, and perhaps very important to the birds because they travel so far without loss of energy, providing possible cues for orientation of their long distance migrations (Yodlowski, Kreithen, and Keeton, 1977). Equally mysterious for us is the electrical sense that some fish use both for object location and for communication (Lissmann, 1958; Hopkins, 1974).

The literature of ethology is full of examples of specialized sensory systems, so highly developed in certain species as to open up a completely new domain of experience for them, to which others, lacking such specializations, are insensitive or even totally blind. Because sensory structures develop innately, according to species-specific genetic instructions, some innate features pervade all perception, by setting limits on what can and cannot be sensed, and by dictating sensitivity to variations within any given sensory domain. Here there is no argument. The disagreement arises about later stages of stimulus processing than mere sensation, phases in which impressions are organized and interpreted, reflected upon, and finally manifest in some appropriate action.

Being so far removed from the chemistry of particular genes, and with so much difficulty in defining which particular structures of an organism are responsible for, say, the transmutations of mental imagery, it is exceptionally difficult to get a grasp on how genes might influence the more subtle aspects of perceptual development. I would like to approach this problem, as ethologists often do, in a comparative vein, as a way of developing my viewpoint that there are elements of truth in both the nativistic and the empiricist approaches to perception. The evidence compels us, I believe, to accept both innate and learned contributions to the development of perception. I want to argue that it is not simply a question of assigning proportions of innateness and learnability, say 80% to 20% in a bird, 60% to 40% in a monkey, and 10% to 90% in our own case. Rather it is a matter of understanding how the two kinds of processes interact, both of them influencing the development of all behavior, affecting all perceptions and all imagery, but in different ways. The diversity of influences will, I believe make more sense to us if we can gain understanding of the differing ways in which each kind of perception serves the organism.

I want to argue that, in animal studies, ethologists, including myself, have *overestimated* the role of innate factors in the development of perception, perhaps in part because the subtler aspects of the perception of another species are so enormously difficult for us to comprehend. By contrast, in dealing with the species with whose perceptual subtleties we are most intimate, our own, I believe that we have *underestimated* the influence of innate factors in perceptual development.

BEHAVIOR OF BACTERIA

In striving to make my case as general as possible I begin with some of the most elementary organisms known to us, namely bacteria. *Escheria coli* hardly seems a promising source of insights to human biology. Yet a distinguished biochemist began a review of the sensory systems of bacteria with a provocative suggestion. A modern molecular biologist, he proposes, might paraphrase the poet Pope with the epithet that "the proper study of mankind is the bacterium" (Koshland, 1977). In so doing, he seeks to shock us into recognizing that, provided we do not forget the differences in the kind of physiological machinery involved, principles of organization of behavior may be very general. He goes on to develop a case that mechanisms involved in bacterial behavior may indeed provide insights into the behavior of more complex organisms, especially because there is a so much better prospect of immediate access to the chemical mechanisms involved. Koshland's scheme of bacterial perception involves an array of stimuli with specific receptors, as well as systems for processing the stimuli, for appraising the relative significance of stimuli that are experienced simultaneously according to something equivalent to a hierarchy of values, and finally for deciding upon an appropriate response. The response decision in this case is very simple, to tumble or not to tumble.

Movements of bacteria by means of their flagella are basically of two types. Either they proceed in a straight line, or they suddenly reorient with a random tumbling movement and go off in a new direction (Delbrück, 1972; Berg, 1975). If conditions are improving, the rate of tumbling goes down. If they get worse, the rate of tumbling increases, so that they gradually congregate in places that are good for bacteria and move away from those that are bad for them. To make a judgment on whether conditions are getting better or worse, a comparison must be made across both space and time, thus requiring a memory. It has been shown that bacteria do in fact compare changing conditions over periods of the order of 20 seconds, a primitive memory to be sure but one whose chemical basis is immediately accessible to study and will probably be understood in a few years (Gould, 1974). Most interesting of all is the discovery that bacteria have a number of specific stimulus receptors. Because the world of a bacterium is primarily chemical, we are concerned with specific chemoreceptors. The present list totals about 20 distinct sensory systems, and no doubt more will be discovered (Adler, 1975; Parkinson, 1975).

How do you know, one might ask, that the bacterium does not simply metabolize the substance and base its response on whether it gets energy and feels good, or whether some damage is done to the cell? That a true sensory system is indeed involved is shown by the clever trick of substituting for a metabolizable sugar another sugar with a similar molecular configuration but which is not metabolizable. It still evokes the same response.

Certain receptors respond to a single compound, others to several. In the latter case the alternative stimuli may vary in the tightness with which they are bound to the receptor. If there is competition between the two of them, that which is tightly bound will displace the loosely bound one, imposing a kind of choice as to which of the two substances will exert most influence on its behavior—which it will pay most attention to, so to speak.

Perhaps the most remarkable of all is the fact that while some bacterial receptors are produced more or less automatically in the course of normal growth, others, such as the ribose and nitrate receptors, are induced and do not arise unless the bacterium has experienced appropriate growth conditions. In other words some responses are innate, while others are modified or induced by environmental conditions. There is also evidence of a capacity for stimulus integration in the interpretation of complex mixtures of chemical stimuli. A solution of both galactose and ribose will be more attractive than one alone, whereas if a repellent is added the attractiveness will be reduced, to a degree that depends on how much of the repellent there is.

It is clear that bacteria can discriminate between compounds that are very similar in structure and can remember having experienced them. While the machinery involved is surely very different from that in higher organisms, it is very likely that induced enzyme formation, such as is known to provide the link between stimulus and response in bacteria, is also involved in neuronal memory mechanisms in higher organisms. It is quite reasonable to suggest that study of the simpler case may throw light on the more complex one (Gould, 1974). If a bacterium's memory seems short, its life is short, and it would have little use for the longer term memory that we find in vertebrates. In other words the special features are adaptive and make perfect sense when viewed in the light of the particular organism's way of lie, its life span, rate of movement, and the kinds of events in its world that are relevant to its survival and reproduction. Interplay between genes and environment gives some plasticity, but only within limits, set by the genetic constitution, in ways that are presumably adaptive.

THE ETHOLOGY OF PERCEPTION

In the following pages these topics are taken up with my favorite subjects for ethological research, namely birds. They are my own first choice for studies of animal learning, infinitely more talented than the overrated white rat (Beach, 1950) and in some ways better subjects for comparative psychologists than monkeys, even when human comparisons are at issue. A recurrent finding of ethological studies on the ways young animals respond to the world around them is that, soon after birth, they often become spontaneously responsive to particular environmental stimuli that are fraught with special biological significance for them. It was on the strength of this kind of observation that the great ethologist, Konrad Lorenz, coined the term "innate release mechanism." Physiological machinery is implied that permits innate recognition of specific stimuli without the prerequisite of previous exposure and interaction with those stimuli that conditioning would require (Lorenz, 1950). Inspired by the Lorenzian viewpoint, Tinbergen and his colleagues undertook the difficult task of specifying exactly what features of these complex, natural situations were in fact controlling behavior, and which were indeed responded to innately (Tinbergen, 1973). There was a natural tendency to focus on the behavior of young animals, whose previous history is more restricted and easier to control than that of adult animals.

One of the classic studies concerned the behavior of the herring gulls living in one of Tinbergen's favorite research sites on the sandy islands off the coast of Holland. He analyzed the stimuli evoking the pecking response of newly hatched gull chicks. The herring gull feeds its young by standing above them with food. The chick pecks at its bill and eventually hits the food. Exhaustive experiments with different kinds of models revealed that the key stimulus properties evoking this pecking response include a particular bill shape, a red patch at the tip, characterized by both color and contrast, bill orientation and movement (Tinbergen, 1953, 1973). Along with these simple aspects, Tinbergen also felt that some "relational" or "configurational" properties are significant to the newborn chick, such as the position of the red spot on the parent's head and beak. A later investigator, Hailman (1967) showed, however, that it is only after a week or so of normal feeding experience that the chicks become responsive to the three-dimensional and configurational properties of the parent's head that were irrelevant to naive chicks. According to Hailman, a week-old chick is already showing signs of developing a complex perceptual "schema" of the parent head, rather than responding just to a piecemeal set of isolated features. By this age a red spot on the forehead was less effective than one on the bill tip, irrespective of its rate of movement. Attempts to train chicks by food reward to respond to novel stimulus features such as changed color preference, while possible, met with difficulty, suggesting biases in the abstraction of features on which the emerging schema was based.

Tinbergen interpreted the herring gull experiments in terms of an ethological "innate release mechanism," presumably evolved in the gull as a special perceptual mechanism to match responsiveness of the pecking chick to co-evolved stimulus features of the parent's bill. As originally conceived, innate release mechanisms implied an innate perceptual "schema" of the stimulus object (Lorenz, 1935, reprinted 1970). It has gradually become evident from ethological experiments of this type that the term "schema," with its "image-like" connotations, is inappropriate. Instead, one typically finds "a mosaic of extremely simple receptor correlates activated by specific key stimuli" (Lorenz, 1970, note 56, p. 375), or "sign stimuli" (Tinbergen, 1948, 1951).

Innate processes are clearly involved. However, one might nevertheless hesitate to invoke "innate knowledge" of the parent's appearance. Yet the particular stimulus features to which the chick is innately responsive are strongly valent in early occurrences of the pecking response, although only simple, piecemeal components of the total situation. As such they must play a significant role in the acquisition of learned responsiveness of the chick to selected aspects of the parent's appearance, and thus in the acquisition of a full, parental "schema."

I believe we can detect here a subtle transition in ethological thinking. It began with an intense preoccupation with innate mechanisms, endowing organisms with responsiveness to a limited number of specific features of things in their external world of special significance to them in their phylogenetic history. Although it has always been recognized by ethologists that innate recognition can subsequently become entangled with learned responses, as studies proceed we have come to realize that the meshing of innate and learned responsiveness is even more elaborate than originally supposed. I now think that young birds learn a great deal, and that much of this learning is guided by what we have called "innate release mechanisms." I would even go further, and suggest that many of the hypothetical "innate release mechanisms" exist primarily not so much to provide the developing perceptions of an animal with hereditarily fixed reference points that are designed to persist throughout life but rather as something more subtle. At least as they occur in birds and mammals, I am now inclined to view them as general guidelines, themselves modifiable by experience. As such, although allowing for great developmental plasticity, they will nevertheless be insidious in their effects on developing perception, sufficient to ensure that developing perceptual organization emerges in an orderly procession.

Developing this type of speculation further, along lines initiated by another great ethologist, William H. Thorpe (1963), perhaps we can entertain the more extreme notion that young birds and mammals have the potential to eventually acquire responsiveness to almost any perceptible feature of a stimulus object. They will nevertheless be prone to react to some features of natural situations first, as though they are endowed with innate salience. In searching for another example that would take such analyses further, we wanted a set of animal behaviors even more radically affected by learning than those already studied. We hoped for something so elaborate that it might conceivably provide analogies with our own behavior, in which we are so very conscious of the overwhelming influence that cultural inheritance has on the development of the perceptions we have of ourselves, of each other, and of the external world. At the same time we needed a situation in which there was also interaction with innate factors that would hopefully exert different influences in different species, giving some access to the genetic mechanisms involved. The subject we chose was the learning of bird song.

It has become clear from research over the past 20 years or so (e.g., Thorpe, 1961; Marler and Mundinger, 1971) that the songs of most and perhaps all oscine birds exhibit an enormously variable and sometimes exquisitely beautiful pattern of sounds that are in some degree learned. It is true of every songbird studied thus far that a male taken from the wild early enough in infancy and reared in social isolation will develop a song that is somewhat abnormal. Species vary widely in the kinds of abnormalities that develop in this so-called Kasper Hauser situation, and I cannot begin to review all of the intricacies of what has now become a complex field of study. Instead, I want to dwell on a particular set of findings that relate directly to the interplay of innate responsiveness and learning in perceptual development.

First, consider the perception of song by an adult, experienced bird. I want to argue, somewhat speculatively, that this may be much more complicated than we have supposed, perhaps as complex in its way as our perceptions of, say, the human face. The evidence is increasing that the intricate variability you find within the song of almost any oscine bird is not a trivial accident, but comprises a set of controlled variations that have significance and meaning to the birds themselves. Take, for example, the well-studied songs of sparrows. It seems to me that an adult's experimentally demonstrable responsiveness to many different subtle features of the acoustic structure of its song could have developed only through learning. Each male differs from all of his neighbors in the fine details of his song. Field experiments in which recorded songs are played back to territorial males reveal that they are acutely responsive to these variations and clearly discriminate between the songs of immediate neighbors and those living further away. They also respond differently depending on exactly where the song originates, showing that they know precisely where the male who has a grace note on the second syllable of his song normally lives (Falls, 1969).

We know that many sparrow songs exhibit local dialects (Marler and Tamura, 1962; Nottebohm, 1969). Again field experiments show that the birds, both male and female, perceive these differences in dialects and respond to them (Verner and Milligan, 1971). The birds also respond to how a song is delivered, whether strongly or weakly, whether the entire pattern of the song is given, or whether the male loses enthusiasm after the first few notes and omits the conclusion of the song.

We can in fact assemble a list of features of bird songs to which adult birds respond, including some that seem to be configurational in nature (Thorpe and Hall-Craggs, 1976). Responsiveness to some, perhaps many, of these features could only have been acquired through learning. If this is true, then the mental imagery that an adult sparrow has of its species' song must be rich and complex.

Can we then ignore *innate* influences in song learning? The answer is clearly no. For one thing there are "universals" in song structure. Although I have stressed the enormous variability of the song patterns of most birds, careful study reveals that this variation is restricted to certain parameters of the song. There are almost always certain features that are more stable. These "universals" are the features on which ornithologists rely in identifying the species of birds, often done more quickly and accurately by voice than by plumage.

There is also direct experimental evidence of innate influences. If we look back into a bird's early life and examine the process by which the song was learned originally, we find an interesting paradox. Most bird songs are learned, yet the learning is selective. Everyone knows that some birds learn sounds not only from their own species but from others as well, as in the mockingbird, the European starling, the lyrebird of Australia, and other famous mimics. But these are exceptions, and it is more general to find that, even in those songbirds in which a Kaspar Hauser male sings a highly abnormal song, a wild male very rarely learns the song of other species.

How do such birds avoid learning the wrong song? In some species this is done through a social mechanism. The young birds learn from an adult of their species with whom they have established a particular social bond, typically the father. They learn whatever sounds he produces, and, because he is likely to be a member of their species, the learning process is canalized so as to preserve a certain set of species-specific characteristics. Having learned the main skeleton of the father's song, they will then often improvise on certain aspects so that individuality comes to mingle with stereotypy.

There are some even more interesting birds in which the

guidance of the learning process, while perhaps potentially influenced by social mechanisms, can also be achieved by a physiological mechanism that is in the possession of the individual bird. An example would be the white-crowned sparrow. Figure 1 diagrams the development of song in young male white-crowned sparrows both under natural conditions and in the face of experimental variation of the acoustic environment. It illustrates three key points.

Suppose that a male is presented with a choice of two songs to learn during the sensitive period for this process, one a song of his own species, the other a song of a close relative living in the same environment. The result is that the young bird focuses attention on the conspecific model and apparently ignores the alien one. The second key point in this reconstruction of the white-crowned sparrow song-learning process is that the song of a Kaspar Hauser, although abnormal in many respects, has some natural features, including the presence of long-sustained whistles. It might be noted that this feature also distinguishes white-crowned song from song sparrow song (Marler and Tamura, 1964; Marler, 1970). The final point of interest is that these normal features are lost if a young male is deafened early in life. The song he develops without the ability to hear his own voice is a noisy tuneless buzz (Konishi, 1965). We infer that the capacity to produce the sustained whistles rests upon an auditory ability and have suggested that the very same ability also provides a simple way of distinguishing between white-crowned song and song sparrow song. In the face of some significant natural choices this ability thus focuses the attention of the young male upon sounds of his species.

These findings give rise to the *modifiable auditory template hypothesis* for selective vocal learning in birds (Marler, 1976). Our notion is that a simple stimulus filter focuses attention on sounds with a particular set of simple properties. The auditory song-filtering function becomes modified in the process of listening to acceptable sounds, so that it is more highly specified. Now it embodies not only additional features of the species' song but also the particular dialect to which the young male has been exposed in his youth. Like many song birds, the young male white-crown learns to sing from memory, only embarking on the gradual transition from subsong to full song after the sensitive period for the first, sensory phase of the learning is completed. The male then behaves as though his next task is to match his vocal output to

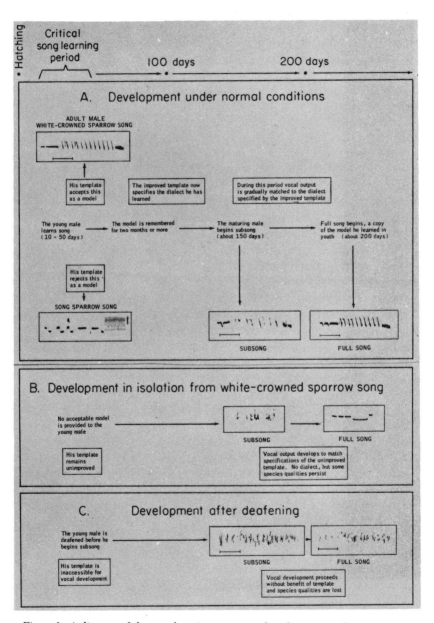

Figure 1. A diagram of the song-learning process in the white-crowned sparrow.

the memory of the particular pattern of sounds to which he was last exposed some weeks or even months earlier, during the sensitive period. It is in this sense that the modified auditory filter becomes used as a kind of template against which the young male matches his song, going through a series of increasingly accurate approximations to the pattern heard in youth until the representation is perfect, apart from those particular features of the song reserved for individual modifications and inventions (Marler and Mundinger, 1971).

I want to concentrate especially on the initial phase of the process in which the young male is first exposed to a medley of sounds from the environment. He apparently rejects some sounds for the process of song learning, accepting only a limited set that have properties diagnostic of conspecific song. As an ethologist I first found myself tempted to invoke an innate release mechanism in the original Lorenzian sense. It seemed natural to view the young male as possessing something equivalent to an innate auditory image, albeit embodying only skeletal features of the song of the species. The young male would use this to select the appropriate model for learning, and then pad out some of the detail through experience. By such a view "innate knowledge of the species song might be an appropriate conception to invoke.

What more can we find out about the nature and extent of that innate knowledge? For a variety of reasons we found the white-crowned sparrow less convenient for pursuing this question further than another sparrow living in the New York area which also learns its songs in a selective fashion. This is the swamp sparrow, having the advantage that it lives in close proximity with the song sparrow, a relative so close that it is placed in the same genus by bird taxonomists, yet whose song is quite different. Almost every young swamp sparrow male has both swamp and song sparrows singing within earshot. How does it avoid learning the wrong song?

The normal songs of males of the two species, although similar in duration, are very different in internal structure (Figure 2). Swamp sparrow song is simple and monotonous, consisting of a slow trill of similar, slurred, liquid notes. That of the song sparrow is more complex, with several parts, consisting of many short notes and a trill near the end. Within these different, relatively stable patterns of singing, both exhibit a great deal of individual variability in the acoustic structure of song "syllables." Swamp sparrows engage in vocal learning, and songs of socially-isolated males are

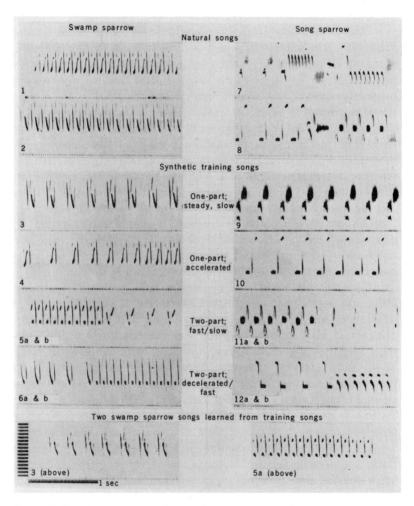

Figure 2. Sound spectrograms of natural song sparrow and swamp sparrow songs and artificial training songs. Natural songs are shown at the top. Syllables from these and others were assembled in synthetic songs, some created from swamp sparrow syllables (e.g., 3–6), some from song sparrow syllables (e.g., 9–12), some in "swamp sparrow-like" patterns (e.g., 3, 4, 9, 10), some in "song sparrow-like" patterns (e.g., 5, 6, 11, 12); syllables from songs 1 and 2 can be seen in songs 3 and 4, syllables from song 8 in songs 10 and 11. Only swamp sparrow syllables were learned. At the bottom are two songs of male swamp sparrows copied from training songs 3 and 5. A 1-sec time marker is given at the bottom left, with a 500-Hz interval frequency scale.

significantly abnormal, with simpler syllables and fewer component parts than usual.

We designed a study to determine whether male swamp sparrows learn selectively when presented with swamp and song sparrow songs, a choice that confronts them in nature (Marler and Peters, 1977). If selective learning occurred, we also wanted to specify some of the acoustic parameters involved in the choice. For this purpose a series of semisynthetic songs was created. Distinctively different sound elements or syllables were edited out from tape recordings of normal local songs of both species. We then spliced these together in a variety of simple temporal patterns, based on features of organization in which normal songs of the two species differ. Some patterns were "swamp sparrow-like." These included sequences of identical syllables at various steady rates. "Song sparrow-like" features included variable rates of delivery of syllable sequences (accelerating, decelerating) and a two-part structure. Our frank expectation was that any choice would be based on these differences in overall patterning of the song.

The great diversity of syllable types from which normal songs are made up led us to doubt the likelihood of significant species differences at the syllabic level. However just as a precaution, each set of 10 temporal patterns was created in duplicate, once with 16 different swamp sparrow syllables, and again with 16 different song sparrow syllables. These 20 "synthetic" songs were then arranged on recording tapes in bouts of each type at normal singing rates. The syllable types had been selected to be sufficiently distinct from one another that if imitation occurred we could determine from which temporal patterns they were selected (Figure 2).

After having been trained with the tapes between 20 and 50 days of age, male swamp sparrows learned to sing from memory, some months after termination of the training. They copied many of the synthetic songs, and every one consisted of swamp sparrow syllables. No song sparrow models were copied. Thus, the male swamp sparrows were extremely selective in their learning, accepting only conspecific syllables for imitation and eschewing song sparrow syllables. The latter were rejected even though they were presented in similar temporal patterns.

The choice was clearly made at the level of components from which the song is constructed, not from the overall pattern of the song. Thus, only one-third of the learned syllables were extracted from one-part songs, the normal swamp sparrow pattern. Two-thirds were from two-part songs, closer to the song sparrow pattern.

While some of the accepted models came from series with a steady rate, the normal swamp sparrow pattern, even more came from accelerated or decelerated series. In the next year we repeated this experiment with male swamp sparrows taken not as young nestlings but as newly laid eggs, which were then foster reared by canaries in the laboratory. The result was identical, showing that the selectivity is innate.

Naturally enough this result has led us to reexamine the fine structure of the syllables of song sparrow and swamp sparrow song very closely. With the aid of a new method of acoustic analysis (Greenewalt, 1968; Staddon et al., in press) we have indeed found what seems to be a consistent difference between them in one quite elementary property that affects the tonal quality. There is a fine frequency modulation or warble in many song sparrow syllables that is absent from those of the swamp sparrow. Our current experiments make use of computer-synthesized patterns of sound in which this feature is varied systematically.

The point to be emphasized here is that the first level of innate responsiveness seems to occur not to the overall pattern of species-specific song but rather to some property of its component parts, a conclusion reminiscent of W. H. Thorpe's (1958) inference from studies of selective song learning in the European chaffinch. This limited degree of innate responsiveness leaves ample scope for the acquisition through learning of sensitivity to other more complicated features involved in special perception of song while sufficing for the simple task of focusing the young bird's attention on a biologically appropriate set of learning models.

In thinking about the physiological mechanisms that must underlie this initial perceptual discrimination, it again seems inappropriate to invoke a complete auditory image of the main features of the song. Rather, what the bird seems to possess are some simple guidelines or signposts for a process of auditory learning.

THE PERCEPTION OF SPEECH

Again you will notice a shift of emphasis, away from innate mental imagery as a major, durable framework for perceptual development, to a more subtle role for innate perceptual predispositions, in providing an ontogenetic basis for adult perceptions that are largely learned. I believe this viewpoint to be valid, even in species that we tend to think of as being so instinct-dominated as birds.

Above all, I feel that this approach provides a suitable basis for thinking about possible innate influences on the development of the incredibly complex patterns of behavior that characterize our own speech. Below some recent research on linguistics is reviewed briefly, dealing with the perception of speech sounds, and the development of responses to speech in very young infants, that suggests to me a quite remarkable convergence with the view I have just set forth of perceptual development in songbirds.

Approaching recent research on the structure of speech sounds as a novice, I was astonished to discover that descriptions of certain physical features of speech patterns of different languages reveal the existence of universals in the properties that define boundaries between functionally distinct patterns of sound. I can best illustrate the results from these comparative vocal "ethograms" by reference to the distinction in many unrelated languages between critical pairs of voiced and unvoiced consonants. I have in mind the property known as "voice-onset-time," a focus of special study because it is one of the few characteristics of speech that can be reliably measured from the frequency/time sound spectrograms on which so many bioacoustic studies are based (Figure 3).

Cross-cultural studies have shown that all languages studied employ voice-onset-time (VOT) as one basis for differentiating voiced and unvoiced consonants in speech (Lisker and Abramson, 1964). Furthermore, when there are VOT boundaries, they always fall in approximately the same places, at one of two locations. Universals have also been found in the patterns of formant onset that differentiate speech sounds produced at different points of articulation in the mouth, labial, alveolar, and velar (e.g., [ba]-[da]-[ga]). There is a long list of other universals (e.g., Greenberg, 1969; Studdert-Kennedy, 1977), but these properties of consonants have the advantage that they are specific and lend themselves to precise analysis as well as computer synthesis.

When such universals are discovered in ethograms of animal behavior, recurring in separate populations of the same species, and contrasting in their uniformity with the more variable properties of vocal dialects, feeding traditions, and other divergent traits of local populations, an ethologist is likely to entertain the possibility of genetic developmental controls.

A further revelation to me was that, although we hear speech sounds as discretely distinct from one another, they are often distributed in actual speech in "graded" continua rather than in

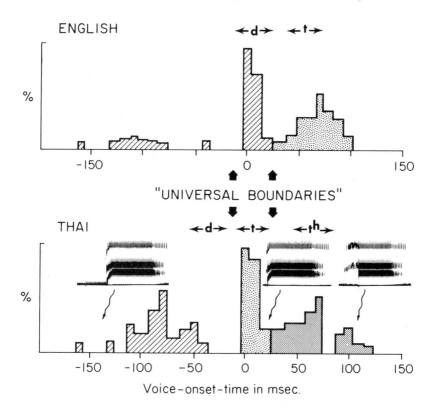

Figure 3. Measurements of speech sounds and their perception: histograms of voice-onset-times in stop consonants in English and Thai. The large arrows indicate the position of the perceived "universal" boundaries. Inserts show three examples of synthetic speech with VOTs of −150 msec, +10 msec, and +100 msec. (After Lisker and Abramson, 1964, and Cutting and Eimas, 1975).

discretely separate categories. For example, histograms of VOTs used in speech reveal that, although the values across a given boundary such as the one between [pa] and [ba] tend to be grouped separately, there are nevertheless intermediate values. Why then are we not more often confused as to precisely which consonant a speaker intends? This implication was not lost on experimental psychologists and led to a series of important studies on responsiveness of adult subjects to graded speech-sound continua, drawn first from natural speech patterns and subsequently created by computer synthesis, with all characteristics under experimental control (Liberman et al., 1957, 1961, 1967).

Such series as the VOT continuum, in 10 or 20 msec steps from [pa] to [ba] to [mba], created by the speech synthesis facilities of the Haskins Laboratories in New Haven have provided the basis for many insights into the mysteries of speech perception. In particular, they have led to evidence for a distinctive mode of perceptual processing that has been dubbed "categorical perception." Although not unique to the perception of speech sounds (e.g., Cutting and Rosner, 1974), nor restricted to the auditory modality (Pastore, 1976), it is especially well exemplified in responses of human subjects to complex acoustic continua such as a VOT series. Asked to label sounds on such a continuum, an English-speaking subject divides them into two groups, labeling one side [pa], the other [ba], with a sharp boundary between that coincides with the trough in VOT productions. This boundary recurs in different languages, although with details that vary consistently from one to another. In some languages, such as Thai, there is a second boundary, around −10-msec VOT, shared by the speech patterns of other cultures as well (Figure 3).

There is another characteristic of so-called "categorical perception" of speech sound continua. If you test adult subjects for the discriminability of sound pairs differing by small increments on the VOT continuum, they display greater sensitivity to variations in the zone of the boundary than to within-category variations (Studdert-Kennedy et al., 1970). As adults we behave as though we are desensitized to within-category variations in this particular property of speech sounds, while being acutely sensitive to small changes at the boundary. This perceptual "quantization" of certain stimulus dimensions contrasts with the "continuous" perception of other kinds of sound properties such as pitch or loudness. Categorical processing thus has the consequence of grouping varied stimuli in classes, imposing a particular kind of order on varying patterns of stimulation. Although virtually unexplored by ethologists, we should seriously entertain the possibility that animals exhibit analogous perceptual phenomena.

Much of what I have described about adult perception of speech could just be thought of as a consequence of the rich perceptual and motor experience of speech that any mature adult brings to bear on a given task of speech-labeling or discrimination. Another set of recent findings, however, suggests that special perceptual predispositions that are independent of prior experience with speech are also involved. A variety of measures of human infant responses to speech sounds, including habituation of a sucking response, heart-rate changes and evoked brain potentials indicate responsiveness to similar boundary values between functionally distinct speech sounds in subjects as young as one month of age or less (Eimas, 1974, 1975; Kuhl, 1976). The early age at which these results are obtained led to the speculation that responsiveness to some of these boundary properties may be innate.

Evidence for an innate component was obtained by Lasky, Syrdal-Lasky, and Klein (1975) in studies of speech perception on four- to six-month-old infants living in a Spanish-speaking environment. There are slight but consistent differences in VOT boundaries in adult production and perception in English and Spanish. These led to the prediction that infants would demonstrate boundary limits different from those obtained by Eimas with children living in English-speaking environments, if these were acquired through infantile experience of speech patterns. The infants proved to be responsive to boundaries in both regions of the VOT continuum that are universals, the so-called "English" and "Thai" boundaries, with no sign that experience of the distributions used in Spanish had affected their speech perception at this age.

Perhaps the nicest study is that of Streeter (1975), conducted in Africa. Infant perception of boundaries along the VOT continuum was studied in children exposed to Kikuyu in infancy. This language has the interesting feature that there is only one labial stop consonant, with a VOT of about − 60 msec. Perhaps as a consequence of exposure to the pattern of usage in their culture, two-month-old infants were responsive to one boundary along the VOT continuum somewhere between 0 and − 30 msec. They seemed more responsive to this boundary than the subjects Eimas had studied in an English-speaking environment, although he reports signs of such a boundary as well. However the Kikuyu-exposed infants, although lacking experience of anything equivalent to a [p], also proved to be responsive to a boundary somewhere between + 10 and + 40 msec, thus resembling infants exposed to English and various other languages. Streeter concluded that, while even at this early age there is evidence of interaction between nature and nurture, some phonetic or acoustic discriminations seem to be universal and innate. Others seem to require previous relevant exposure.

Further studies have demonstrated responsiveness in infants between one and six months of age to the variations in second- and third-formant transitions in synthetic speech patterns that establish boundaries between the different articulation points distinguishing labial, alveolar, and velar stop consonants (e.g., [ba], [da], and [ga]). Infants also seem responsive to differences between vowel sounds.

The potential lability of predispositions that human infants may bring to segmentation of speech sound continua is also clear. The [ra]/[la] distinction that Japanese adults find so difficult, un-employed in Japanese, is probably easier for infants, although only American subjects have been tested thus far (Eimas, 1975). It is self-evident that the features of speech sounds on which learned responsiveness in adulthood is based are much more complex than those of infants, with more redundancy, perhaps involving config-urational features rather than simple properties, and sometimes so changed that the effective stimulus set no longer contains those that match the original predispositions of infancy. Nevertheless, the latter must surely play a significant ontogenetic role in setting the trajectory for learning to respond to more elaborate arrays of abstracted features.

A hint of the rich developmental possibilities derives from a study by Kuhl and Miller (1975). The formant patterns that dis-tinguish different vowel sounds are complicated by variations in the fundamental frequency of different voices, as between men and women. Such variations are likely to be a serious distraction for an infant learning to analyze speech. Given the importance of vowel coding in speech, we might expect a predisposition to focus more strongly on formant patterns than on pitch in early responses. By independently varying the two features in sounds presented to infants, Kuhl and Miller (1975) were able to show that variations in formant pattern are indeed more salient or arresting for human infants than are variations in pitch. This is not to say that they are unresponsive to pitch variations—far from it. However the salience of pitch changes to the infant is less than that of variations in vowel patterns, thus imposing some order on the process of learning to extract different features from the complex array of stimuli that speech sounds present.

These findings about speech patterns, and speech perception in adults and infants, seem to me of particular interest to biologists (Table 1). They show that human babies bring some well-defined

Table 1.   Aspects of speech behavior relevant to ethology

1.   There are cross-cultural universals in acoustic properties defining boundaries between functionally distinct speech sounds.
2.   Some functionally distinct speech sounds are not discretely separated but are connected by a continuous series of graded intermediates.
3.   Adults process graded speech-sound-continua "categorically," by reference to boundaries, rather than "continuously."
4.   Prespeech infants are sensitive to some of these same "universal boundaries."

perceptual predispositions to the task of developing responsiveness to the complex of sound properties that speech represents. Some seem to be innately manifest in initial encounters, developing without prior experience of the stimuli involved.

Although the evidence for innate contributions to the development of speech perception is strong, we are hardly tempted to view such influences as developmental instructions for designing infants as human automata, as might have been the case in some circles, given similar results for birds. It seems more natural in the human case to think of these perceptual predispositions as helping the infant to learn, by providing initial instructions that set the trajectory for development of learned responsiveness to a more elaborate array of abstracted features. Eventually information about these features becomes embodied in the mental images or "schemata" invoked by many psychologists in conceptualizing the development of mature human perceptions of complex stimuli (Marler, 1977). If such inferences follow in the human case, parallel arguments for birdsong perception are no less compelling.

CONCLUSION

Thus, I am inclined to picture these innate influences, whether on the development of the perception of speech and the actions of speaking in our own species, on birds' development of the perception of song, and singing, or on the development of the visual perception that a gull has of its parents, in terms of something less complex than innate mental imagery, or *innate knowledge*. The innate instructions are relatively remote from the adult capability about which knowledge is being assumed. To serve as a basis for the development of perceptions that are then employed, template fashion, to guide the development of motor activities through sensory feedback, innate influences must be modified, often drastically, through subsequent learning. Their primary role must surely be to influence the direction that such learning is to take. Relatively simple mechanisms are likely to be involved, such as those that imbue some stimuli with high initial valence, or reduce rates of habituation to certain stimuli rather than others, or heighten responsiveness to stimulus variations at certain zones of a stimulus continuum.

Such influences, often with properties that will vary from species to species, provide a developmental basis for the diverse behavior of organisms of different species. As young grow up they will perceive their external worlds differently because their perceptions guide them in different directions. They will thus be led to behave differently as adults, as different in their ethology as in their structure and appearance. It behooves us not to neglect such effects if we wish to comprehend the machinery of behavioral development.

It is my belief that the prospect of further understanding is improved if we hesitate before leaping to the inference from studies of innate perceptions that there must be "innate knowledge." We should first explore the possibility that what is innate is something less than complete perceptual imagery. I have shown that there are precedents for relatively simple innate perceptual biases, which, exerted in youth, have powerful consequences for the ordering of subsequent development within that particular sensory modality. What follows ensures some sharing of common principles of perceptual operations. A degree of rule sharing is particularly important in the perceptual analysis of stimuli involved in communication. The kind of mechanisms suggested here can help to achieve this, while still leaving latitude for the enormously rich modification of perceptions and actions that we treasure in our own species, and which, we are now beginning to realize, are more common among animals than previously supposed.

The neo-nativist position developed here can also help to resolve the controversy over the role of "innate knowledge" in the development of language and, in particular, in the universal use in all human languages of a certain class of transformational grammars (Lenneberg, 1967). Chomsky's argument that the apparent universality of this extremely special system of grammar implies an innate component in what he dubs the "language acquisition device" seems to me a compelling one (Chomsky, 1971). Yet it is not clear what minimal innate influence one most postulate for this universality to be achieved.

Students of the language learning of children stress the role of semantic use in the shaping of language. Language grows out of a rich background of mutual nonverbal exchange and social interaction with others, as well as through experience of the external world (e.g., Greenfield and Smith, 1976; Lewis and Rosenblum, 1977). While of great intrinsic interest, much of this work addresses other issues than that of universal grammar. Even when the question of universals *is* the focus (e.g., Putnam, 1971; Inhelder, Sinclair, and Bovet, 1974), the relevance to Chomsky's position is often unclear, perhaps because he has never, to my knowledge, set down an explicitly developmental view of the "innateness hypothesis." Might it not suffice to achieve universality if no more than a very few early steps in cognitive and language development were innate? What of the possiblities of some simple rules for symbolic behavior, or for apposition of words with different semantic functions such as object labeling and action identification or with different social functions such as "commanding" or "declaring" (Bates et al., 1977)? Following in the wake of such early cognitive predispositions, some of the detail of grammar that evolves subsequently through experimentation with efficient and inefficient communicative use might be sufficiently constrained by these early steps that common rules of deep structure would be adhered to. Growing abilities to achieve coherence and to eliminate redundancy in the employment of linguistic rules might also play a part. In other words, a given universal rule might arise "from the interaction of other more basic properties of the language faculty" (Chomsky, 1976, p. 22). If something like this were to prove to be the case, thus confirming Chomsky's hypothesis, I would still question whether "innate knowledge" is the proper term to apply to that which is innate in the language development process, even though innate influences may be critical in determining the nature and shape of the linguistic "knowledge" of adults.

Thus, I feel we have in prospect a resolution of the ancient nativist-empiricist controversy. The solution I envisage makes sense in biological terms because it involves processes that are in principle relatively simple, widespread, and susceptible to investigation. If they are brought into play at the appropriate time in an organism's life they can have just as profound consequences for the developing perceptions of a species as the infinitely more elaborate and, I feel, more improbable linguistic and philosophical notions of complete, innate knowledge.

ACKNOWLEDGMENTS

The author is indebted to Professor James Gould for productive discussion of the issues raised in this chapter. Professors Max Delbrück, James Gould, Donald Griffin, Donald Kroodsma, and Fernando Nottebohm made valuable suggestions for improving the manuscript. I am grateful to Professor Robert Long for comments and advice on philosophical aspects of "innate knowledge."

REFERENCES

Adler, J. 1975. Chemotaxis in bacteria. Ann. Rev. Biochem. 44:341–356.

Barlow, H. B. 1975. Visual experience and cortical development. Nature 258:199–204.

Bates, E., Benigni, L., Bretherton, I., Camaioni, L., and Volterra, V. 1977. From gesture to the first word: On cognitive and social prerequisites. In M. Lewis and L. A. Rosenblum (eds.), Interaction, Conversation and the Development of Language. Wiley Interscience, New York.

Beach, F. A. 1950. The shark was a Boojum. Am. Psychologist 5:115–124.

Berg, H. C. 1975. Chemotaxis in bacteria. Ann. Rev. Biophys. Bioeng. 4:119–136.

Berkeley, G. 1709. Essay Towards a New Theory of Vision. E. P. Dutton & Co., London.

Berkeley, G. 1710. A Treatise Concerning the Principles of Human Knowledge. Tonson, London.

Blakemore, C. 1977. Mechanics of the Mind. Cambridge University Press, Cambridge, Eng.

Chomsky, N. 1971. Recent contributions to the theory of innate ideas. In J. R. Searle (ed.), The Philosophy of Language. Oxford University Press, Oxford.

Chomsky, N. 1976. On the biological basis of language capacities. In R. W. Rieber (ed.), The Neuropsychology of Language. Plenum Press, New York.

Cutting, J. E., and Eimas, P. D. 1975. Phonetic feature analyzers and the processing of speech in infants. In J. F. Kavanagh and J. E. Cutting (eds.), The Role of Speech in Language. MIT Press, Cambridge, Mass.

Cutting, J. E., and Rosner, B. 1974. Categories and boundaries in speech and music. Percept. Psychophys. 16:564–570.

Delbrück, M. 1972. Signal transducers: Terra incognita of molecular biology. Ange. Chem. 11:1–6.

Eimas, P. D. 1974. Auditory and linguistic processing of the cues for place of articulation by infants. Percept. Psychophys. 16:513–521.

Eimas, P. D. 1975. Speech perception in early infancy. In L. B. Cohen and P. Salapatek (eds.), Infant Perception: From Sensation to Cognition, Vol. II. Academic Press, New York.

Falls, J. B. 1969. Function of territorial song in the white-throated sparrow. In R. A. Hinde (ed.), Bird Vocalizations. Cambridge University Press, Cambridge, Eng.

Gould, J. L. 1974. Genetics and molecular ethology. Z. Tierpsychol. 36:267–292.

Greenberg, J. H. 1969. Language universals: A research frontier. Science 166:473–478.

Greenewalt, C. H. 1968. Birdsong, Acoustics and Physiology. Smithsonian Institution Press, Washington, D.C.

Greenfield, P. M., and Smith, J. H. 1976. The Structure of Communication in Early Language Development. Academic Press, New York.

Gregory, R. L. 1974. Concepts and Mechanisms of Perception. Charles Scribner's Sons, New York.

Griffin, D. R. 1974. Listening in the Dark. Dover Publications, New York.

Grobstein, P., and Chow, K. L. 1976. Receptive field organization in the mammalian visual cortex: The role of individual experience in development. In G. Gottlieb (ed.), Neural and Behavioral Specificity. Academic Press, New York.

Hailman, J. P. 1967. Ontogeny of an instinct. Behavior Suppl. 15:1–159.

Herrnstein, R. J., Loveland, D. H., and Cable, C. 1976. Natural concepts in pigeons. J. Exp. Psychol., Animal Behavior Processes 2:285–302.

Hochberg, J. E. 1962. Nativism and empiricism in perception. In L. Postman (ed.), Psychology in the Making. Knopf, New York.

Hopkins, C. 1974. Electric communication in fish. Am. Sci. 62:426–437.

Hubel, D., and Wiesel, T. 1963. Receptive fields of cells in striate cortex of very young, visually inexperienced kittens. J. Neurophys. 26:994–1002.

Inhelder, B., Sinclair, H., and Bovet, M. 1974. Learning and the Development of Cognition. Harvard University Press, Cambridge, Mass.

Konishi, M. 1965. The role of auditory feedback in the control of vocalization in the white-crowned sparrow. Z. Tierpsychol. 22:770–783.

Koshland, D. E. 1977. A response regulator model in a simple sensory system. Science 196:1055–1063.

Kuhl, P. K. 1976. Speech perception in early infancy: The acquisition of speech-sound categories. In S. K. Hirsh, D. H. Eldredge, I. J. Hirsh, and S. R. Silverman (eds.), Hearing and Davis: Essays Honoring Hallowell Davis. Washington University Press, St. Louis.

Kuhl, P. K., and Miller, J. D. 1975. Speech perception in early infancy: Discrimination of speech-sound categories. J. Acoust. Soc. Am. Suppl. 1:58, 56.

Lasky, R., Syrdal-Lasky, A., and Klein, R. 1975. VOT discrimination by four-to-six-month-old infants from Spanish environments. J. Exp. Child Psychol. 20:215–225.

Lenneberg, E. H. 1967. The Biological Foundations of Language. John Wiley & Sons, New York.

Lewis, M., and Rosenblum, L. A., (eds.). 1977. Interaction, Conversation and the Development of Language. Wiley Interscience, New York.

Liberman, A. M., Cooper, F. S., Shankweiler, D., and Studdert-Kennedy, M. 1967. Perception of the speech code. Psychol. Rev. 74:431–461.

Liberman, A. M., Harris, K. S., Hoffman, H. S., and Griffith, B. C. 1957. The discrimination of speech sounds within and across phoneme boundaries. J. Exp. Psychol. 54:358–368.

Liberman, A. M., Harris, K. S., Kinney, J. A., and Lane, H. 1961. The discrimination of relative-onset time of the components of certain speech and nonspeech patterns. J. Exp. Psychol. 61:379–388.

Lisker, L., and Abramson, A. S. 1964. A cross-language study of voicing in initial stops: Acoustical measurements. Word 20:384–422.

Lissmann, H. W. 1958. On the function and evolution of electric organs in fish. J. Exp. Biol. 35:156–191.

Locke, J. 1690. An Essay Concerning Human Understanding. London: Printed for Thomas Basset.

Lorenz, K. 1950. The comparative method in studying innate behavior patterns. Symp. Soc. Exp. Biol. 4:221–268.

Lorenz, K. 1970. Studies in Animal and Human Behavior. Vol. I. Harvard University Press, Cambridge, Mass.

Marler, P. 1970. A comparative approach to vocal learning: Song development in white-crowned sparrows. J. Comp. Physiol. Psychol. 71:1–25.

Marler, P. 1976. Sensory templates in species-specific behavior. In J. Fentress (ed.), Simpler Networks: An Approach to Patterned Behavior and Its Foundations. Sinauer Assoc., New York.

Marler, P. 1977. Development and learning of recognition systems. In T. H. Bullock (ed.), Recognition of Complex Acoustic Signals. Dahlem Konferenzen, Berlin.

Marler, P., and Mundinger, P. 1971. Vocal learning in birds. In H. Moltz (ed.), Ontogeny of Vertebrate Behavior. Academic Press, New York.

Marler, P., and Peters, S. 1977. Selective vocal learning in a sparrow. Science 198:519–521.

Marler, P., and Tamura, M. 1962. Song variation in three populations of white-crowned sparrow. Condor 64:368–377.

Marler, P., and Tamura, M. 1964. Culturally transmitted patterns of vocal behavior in sparrows. Science 146:1483–1486.

Nottebohm, F. 1969. The song of the chingolo, *Zonotrichia capensis*, in Argentina: Description and evaluation of a system of dialects. Condor 71:299–315.

Parkinson, J. D. 1975. Genetics of chemotactic behavior in bacteria. Cell 4:183–188.

Pastore, R. E. 1976. Categorical perception: A critical re-evaluation. In S. K. Hirsh, D. H. Eldredge, I. J. Hirsh, and S. R. Silverman (eds.), Hearing and Davis: Essays Honoring Hallowell Davis. Washington University Press, St. Louis.

Putnam, H. 1971. The "innateness hypothesis" and explanatory models in linguistics. In J. R. Searle (ed.), The Philosophy of Language. Oxford University Press, Oxford.

Ratliff, F. 1965. Mach Bands. Holden-Day, San Francisco.

Schrodinger, E. 1945. What Is Life? Cambridge University Press, Cambridge, Eng.; Macmillan Co., New York.

Searle, J. R. 1971. The Philosophy of Language. Oxford University Press, Oxford.

Staddon, J. E. R., McGeorge, L. W., Bruce, R. A., and Klein, F. A simple method for the rapid analysis of animal sounds. Z. Tierpsychol. In press.

Streeter, L. A. 1975. Language perception of 2-month old infants shows effects of both innate mechanisms and experience. Nature 259:39–41.

Studdert-Kennedy, M. 1977. Universals in phonetic structure and their role in linguistic communication. In T. H. Bullock (ed.), Recognition of Complex Acoustic Signals. Dahlem Konferenzen, Berlin.

Studdert-Kennedy, M., Liberman, A. M., Harris, K. S., and Cooper, F. S.

1970. Motor theory of speech perception: A reply to Lane's critical review. Psychol. Rev. 77:234–249.

Thorpe, W. H. 1958. The learning of song patterns by birds, with especial reference to the song of the chaffinch *Fringilla coelebs*. Ibis 100:535–570.

Thorpe, W. H. 1961. Bird Song: The Biology of Vocal Communication and Expression in Birds. Cambridge University Press, Cambridge, Eng.

Thorpe, W. H. 1963. Learning and Instinct in Animals. Harvard University Press, Cambridge, Mass.

Thorpe, W. H., and Hall-Craggs, J. 1976. Sound production and perception in birds as related to the general principles of pattern perception. In P. P. G. Bateson and R. A. Hinde (eds.), Growing Points in Ethology. Cambridge University Press, Cambridge, Eng.

Tinbergen, N. 1948. Social releasers and the experimental method required for their study. Wilson Bull. 60:6–51.

Tinbergen, N. 1951. The Study of Instinct. Clarendon Press, Oxford.

Tinbergen, N. 1953. The Herring Gull's World. Collins, London.

Tinbergen, N. 1973. The Animal in its World. Vols. I and II. Harvard University Press, Cambridge, Mass.

Verner, J., and Milligan, M. 1971. Responses of male white-crowned sparrows to playback of recorded song. Condor 73:56–64.

von Frisch, K. 1967. The Dance Language and Orientation of Bees. Belknap Press, Harvard University Press, Cambridge, Mass.

Yodlowski, M. L., Kreithen, M. L., and Keeton, W. T. 1977. Detection of atmospheric infrasound by homing pigeons. Nature 265:725–726.

chapter 5
Mind from Matter ??

Max Delbrück

Division of Biology
California Institute of Technology
Pasadena, California

> "The world has visibly been recreated
> in the night.
> Mornings of creation, I call them.
> In the midst of these works
> of a creative energy recently active,
> While the sun is rising with more than
> usual splendor,
> I look back . . . for the era of this creation,
> Not into the night, but to a dawn
> For which no man ever rose early enough."
> Henry David Thoreau

PREFACE: SCHROEDINGER'S BOOK

In their invitation to the speakers the organizers of the XIIIth Nobel Conference made reference to Schroedinger's book *What Is Life? The Physical Aspects of the Living Cell* (1945), which appeared some 30 years ago, just as World War II ended. Physicists returning from war work and many freshmen just entering college were attracted to biology by this book. It may, therefore, be fitting to preface this chapter with a short consideration of it.

At the outset Schroedinger proposed to discuss "the large and important" question: "How can the events *in space and time* which take place within the spatial boundary of the living organism be accounted for by physics and chemistry?"

This is a broad question, embracing the problem content of many branches of the biological sciences. Biologists with a romantic reverence for the powers of modern physics might have been led to

141

expect great revelations from the book. It turns out, however, that the author's aims were much more modest. It was not at all the "how" of the accounting in which he was interested, but *whether* physics and chemistry *would be* able to give a *complete* account. This is not as clear-cut a question as it might seem to be because the terms physics and chemistry are ill-defined when applied not to the present but to some future content of these sciences. It is also a question whose answer involves phophecy, and in this respect it is interesting indeed to compare the prophecy with what in fact happened. In his main discussion the author climbed down to a still more modest ground and defined as his aim the establishment of this point of view:

> The obvious inability of present-day physics and chemistry to account for such events (i.e., the peculiar things living cells do) is no reason at all for doubting that they can be accounted for by those sciences.

The argument ran as follows. In physics we have learned to describe the well-determined behavior of systems by two types of law: 1) either the systems are made up of large solid bodies, like clocks, and follow dynamic laws unperturbed by molecular heat motion, or 2) they are made up of very large numbers of statistically independent elements and observation concerns the average behavior of these elements. We are then dealing with statistical laws that seem to be accurate laws because the numbers of particles involved are so large that statistical fluctuations are negligibly small. The diffusion law and the mass action law are examples of this type of law in which "order comes from disorder." In either of these two types of well-determined behavior it is essential that individual molecules have no measurable effect on observable quantities.

Being well acquainted with this situation, the naive physicist would expect that the living cell, exhibiting a well-determined behavior, is constructed on the same principles, i.e., that it functions like a clock or like a thermal machine, or some mixture of the two. In any event he would expect that it is well protected from being thrown off balance by statistical fluctuations. Therefore, if the cell contains molecular species with an important role in its makeup, he would expect each such molecular species to be represented by a very large number of individual molecules. In these expectations the naive physicist finds himself flatly contradicted by the facts of life. Far from being a mass action system, either in the dynamic or in the thermodynamic sense, the living cell is controlled by genes, which are present in single or double copy and which are handed on from cell to cell by an unknown mechanism of reproduction.

This outline of the naive physicist's dilemma is followed by an exposition of some of the elements of genetics, of the results of radiation genetics, and of other data bearing on the mechanism of mutation and on the stability of the genes. The discussion leads up to the concept of the gene as an "aperiodic crystal." The genes are given this startling name rather than the current name "macromolecule," because the author wished to emphasize, first, that the stability of the genes is of the same kind as the stability of crystals in which the constituent atoms are held together by valency bonds and, second, that these crystals are of a new kind, not previously investigated by physicists, who concentrated their attention on periodic crystals.

There was nothing new in this exposition, to which the larger part of the book is devoted, and biological readers no doubt were inclined to skip it.

Schroedinger laid great stress on the fact that the stability of the genes is a quantum mechanical stability. We knew then that the stability of atoms and molecules is deeply rooted in quantum mechanics, and our general understanding of molecular stability was of help when the stability of genes was investigated. This indebtedness to quantum mechanics never was a private debt of biology. It is part of the general debt of chemistry to quantum mechanics. In Schroedinger's opinion the dependence of the well-determined behavior of living organisms on this new and little studied physical structure, the "aperiodic crystal," explained the obvious inability of the then current physics and chemistry to account for this behavior. In his opinion this inability is similar to the inability of a man who knows how a steam engine works and who is confronted with an electric generator. Both of these contraptions contain iron and copper, but unless he has studied electrical phenomena the steam engineer will be quite unable to understand the working of the generator. He might be tempted to believe that the laws of physics break down when applied to the generator, whereas in reality he is merely confronted with the behavior of matter under a new set of conditions that he has not yet analyzed.

This bold analogy seemed very suggestive to many readers. But was it a valid analogy? One must note that it is based in large measure on a substitution of terms. For the term "macromolecule"

Schroedinger substituted the term "aperiodic crystal," the latter term implying a new and unexplored state of matter. In retrospect, we should say Schroedinger was in large measure right: most of the insights of molecular biology of the intervening decades have indeed come from the studies of the properties of macromolecules, i.e., proteins and nucleic acids.

After his profession of faith in the physical nature of the workings of the cell, Schroedinger attempted a further characterization of life as a class of systems that produce order from order, in contradistinction to the statistical mechanism in which the orderliness of the observed phenomena is the result of the disorder of the molecules. The clockwork is also based on the order-from-order principle. The basic similarity between the clock and the living cell is seen in the fact that both depend on the solidity of the controlling parts: the clockwork relies on massive manmade solids, in which the disordering tendencies are kept at bay by operating with solids of large mass, while in the living cell the decay to thermodynamic equilibrium is evaded by continually drawing on sources of negative entropy from its environment.

While it is of course correct to say that life processes depend on the availability of negative entropy, such a statement is at best a partial description of the tricks of life. It fits any system of enzymes acting on their substrates. Translated into the usual terminology it reads: The living cell employs specific enzymes, and these enzymes promote selectively some of the reactions that involve an increase in free energy in a subsystem: the cell, or the organism.

The author did not return in his later discussion to the problem of how the cell gets around the statistical fluctuations—a disappointing, but in retrospect not a surprising, omission. At the beginning of the book the statistical fluctuations were represented as an insurmountable obstacle to the physical understanding of the cell, but later this difficulty seems forgotten. How does the cell manage to produce just one replica of each gene at each cell division? Are we not dealing with small numbers of molecules, and should we not expect large fluctuations? In the 1940s the knowledge of cellular processes was quite insufficient to make a discussion of these difficulties convincing. Indeed, even today, our understanding of the regulation of DNA replication is very incomplete. At the time the book was written any statement about the physical nature of cellular organization would have been premature.

Schroedinger delimited his stand further by stating that, in his opinion, "and contrary to the opinion upheld in some quarters, *quantum indeterminacy* plays no biologically relevant role [in the cell]." The opinions to which he was referring were presumably those of Niels Bohr and Pascual Jordan. Bohr's argument is returned to in this chapter.

The book contained a brief epilogue on "Determinism and Free Will." Having declared his belief that the events in the cell, if not strictly deterministic, are at least statistico-deterministic, Schroedinger attempted in this epilogue to bring this belief into harmony with the "immediate experience" of a free will, i.e., with the experience of directing the motions of one's body, of which motions one foresees the effects. How can the conscious "I" direct something whose course is set by the laws of nature? Schroedinger thought that the only solution to this dilemma lies in the assumption that the "I" and the laws of nature are one and the same thing. It must then be assumed that the "I" of immediate experience is a singular of which the plural is unknown. He drew on the scholars of the Vedanta and on other sources to support the view that the plurality of consciousness is an illusion.

I believe that modern biology is capable of throwing light on this problem in a less mystical and more concrete form. This will be the main thrust of my talk.

The book attracted many readers, because of its title, because of the very high respect in which its author was held and because of its clear, simple, and forceful style. It had an inspiring influence by acting as a focus of attention for both physicists and biologists. Physicists were interested in the book because to some extent it answered their curiosity about biology in a language they could understand. Biologists found it hard to appreciate the dilemma of the "naive physicist" outlined at the beginning of the book. Physicists and biologists who were not familiar with Bohr's subtle complementarity argument were inclined to take the physical nature of the cellular processes for granted at the outset, and were perhaps dissatisfied because Schroedinger did not advance our understanding of cellular mechanisms in any specific respect. However, Schroedinger's discussion of the types of laws of nature exerted a clarifying influence on biological thinking.

STATEMENT OF THE PROBLEM

Turning now to the main topic of this chapter, an entry from one of Søren Kierkegaard's diaries from 1846 is an appropriate beginning. It reads as follows:

...That a man should simply and profoundly say that he cannot understand how consciousness comes into existence—is perfectly natural. But that a man should glue his eye to a microscope and stare and stare and stare—and still not be able to see how it happens—is ridiculous, and it is particularly ridiculous when it is supposed to be serious. . . . If the natural sciences had been developed in Socrates' day as they are now, all the sophists would have been scientists. One would have hung a microscope outside his shop in order to attract custom, and then would have had a sign painted saying: "Learn and see through a giant microscope how a man thinks" (and on reading the advertisement Socrates would have said: "that is how men who do not think behave"). (Hong and Hong, 1975).

In this chapter I wish to propose, and to propose seriously, doing this ridiculous thing; to "look through the microscope" in an attempt to understand how consciousness,[1] or more generally, how mind comes into existence, and, with mind, how language, how the notion of truth, and how logic, mathematics, the sciences have come into the world.

Ridiculous or not, to look for the origin of mind is no longer an idle question. It has become an approachable, natural, indeed an unavoidable, question.

COSMOLOGY: BEGINNING OF LIFE ON PLANET EARTH

We know enough about the history of the universe—the galaxies, the stars, the planets, our planet Earth—to be certain that the earth started without life, forming its earliest rocks 3.8 billion years ago (Moorbath, O'Nions, and Pankhurst, 1973). We also know that life on Earth today is unmistakably monophyletic: it is based universally on nucleic acids and proteins; it universally involves the same genetic code; each amino acid is coded for by the same triplet of bases in the nucleic acids; the ribosomes, transfer RNAs and charging enzymes are manifestly related; the proteins in their amino acid sequences and in their folding patterns clearly display their descent from common ancestors. The ancestry can be traced back by the pre-Cambrian geological record to at least 3.1 billion years ago (Schopf and Barghoorn, 1967), not long after the formation of the oldest rocks. Life cannot have flown in from outer space: no life exists (nor ever did) on the other planets, and none would survive the exposure to cosmic rays during the journey

---

[1]Consciousness is a term too variously used. A very enlightening discussion of the term *consciousness*, narrowly defined, can be found in Jaynes (1976).

from the far reaches of the universe. The cosmic ray intensity in interstellar space (according to R. Vogt, personal communication) is at least 100 times higher than that assumed by Schklovsky and Sagan (1966). What is actually formed in outer space, by radiation, is the "soup" from which organic molecules are easily formed, i.e., the molecules H_2O, NH_3, CH_2O, $HC \equiv N$, $HC \equiv C - C \equiv N$.

PROKARYOTES: BEGINNINGS OF PERCEPTION

Thus, there is a clear case for the transition on Earth from no-life to life. How this happened is a fundamental, perhaps *the* fundamental question of biology. The difficulty is not that we cannot dream up schemes of how it might have happened. Hundreds of possible schemes can and have been proposed. The difficulty lies in the lack of data: there is *no* geological record of "prebiotic evolution." On the contrary, there is an immense gap between all present-day life and no-life, a gap that has widened enormously during the last decades as the essential unity and immense complexity of all present-day molecular genetics has become apparent.

The unity and continuity of life on Earth is manifest in its anatomy, down to the molecular anatomy. It is manifest, equally, in its "psychic" aspects. The beginnings of perception are clearly present in microorganisms: in their adaptive behavior, permitting the organisms to detect and evaluate signals from the environment and to respond to them appropriately. Is the behavior we see in these situations strictly deterministic? Does it involve elements of decision in situations of conflict? Suppose a bacterium is subjected to a chemotactic stimulus, an attractant, or a repellant? How will it decide how to respond? By a fixed formula? By reference to past experience? The answers to these questions are known: the responses of microorganisms to signals from the environment occur at two levels:

1. The swimming pattern is a stop-go pattern. During the "stop" random rotational diffusion determines in a stochastic way the next direction of "go;" the duration of "go," and thereby the length of the "go" section is determined by the signal. The result is a "random walk," a biased random walk, biased in the direction towards higher concentration of an attractant, toward lower concentration of a repellant.

2. The detection of the signal involves an adaptation mechanism, quite similar to the light and dark adaptation of our eyes. The

organism sets its level of detection in accordance with the average intensity of the stimulus (concentration of chemical, intensity of light, etc.).

Thus we have in bacteria all three components: deterministic response, stochastic response, and response based on (immediate) past experience.

RECONSTRUCTION OF THE TREE OF LIFE
FROM PALAEONTOLOGY AND COMPARATIVE ANATOMY

It is a long way from these primitive "decisions" to the machinery involved in the decision of John Doe to propose marriage to his girlfriend, Susie. John responds to Susie's charm with his emotional brain (the midbrain) and his thinking brain (the cortex). His decision is influenced by his genes, by his imprinting in earliest childhood, by his identity as it detached itself from his mother, by his moral upbringing, by his economic circumstances, by his surging sex drive, and so on.

Let us trace this long way to get a feeling for its continuity. How do we reconstruct the tree of life? How do we identify the common ancestors of the presently living forms? In the first place by digging and dating, or by palaeontology, i.e., finding remains that have survived the ages (e.g., bones of vertebrates and hard parts of invertebrates, impressions, petrified structures). We date the specimens by a variety of physical techniques and try to establish their relationships to each other and to presently living forms. In some cases procedures of this kind can take us back very far, as with some sulfur deposits, the isotope ratios of which indicate a biological action of sulfate-reducing organisms 10^9 years ago. On the whole, palaeontological methods are the *only* ones that give us direct evidence of our ancestors. However, the evidence from comparative anatomy, physiology, and biochemistry of living forms is much richer. The methods in these fields tell us who is related to whom and how closely. They permit us to infer some properties of *actual common ancestors* but not of sidelines that became extinct: no amount of study of present forms would permit us to infer dinosaurs.

Most far reaching of the comparative approaches is the comparative study of informational molecules (Orgel, 1973): proteins and nucleic acid. Progress is being made very rapidly with the advent of efficient methods of determining sequences of amino acids in proteins, bases in nucleic acids, and the folding patterns of proteins.

EVOLUTION OF ENERGY, METABOLISM, AND TRANSITION TO EUKARYOTES

What is emerging from these studies is an outline of the family tree of life-styles among the prokaryotes, especially the family tree of their ways and means to secure the basic element of their chemical energy currency, adenosine triphosphate (ATP): from fermenters to sulfate reducers to photosynthetic mechanisms, the latter culminating in the liberation of O_2 into the atmosphere and the O_2 in turn permitting the evolution of oxygen respiration (Dickerson, Timkovich, and Almassy, 1976).

It is believed that these advanced methods of efficient energy metabolism paved the way to the next big leap: the evolution of eukaryotes—cells with genomes too large to be handled by a single chromosome and therefore requiring the elaborate maneuver of mitosis, the ordered distribution of several chromosomes to daughter cells. Along with this elaboration we find the occurrence of alternating haploid and diploid phases of the life cycle, strict bisexuality, and meiosis. Here again we have a big gap: all of the adaptations just mentioned seem to appear in the evolutionary tree without traces of transitional forms. In recent years it has been found that even the most primitive fungi have a meiosis that follows the strict rules long known from higher organisms: two meiotic divisions with suppression of one cycle of chromosome replication, crossing over in the four-strand stage, and synapsis of the homologous chromosomes with formation of a well-defined elaborate synaptinemal complex.

Here, as in the prebiotic stage, there must have been a vast number of alternative lines that died out without leaving a trace in the progeny, that did not survive the rigors of natural selection. The present eukaryote cells may actually constitute barnyards full of domesticated creatures, their organelle for oxidative phosphorylation (the mitochondrion) having been derived from a symbiotic relationship with a bacterium, possibly paracoccus, while the green plants' organelle for photosynthesis (the chloroplast) may be derived from certain photosynthetic bacteria, specifically from a recently discovered species which contains both chlorophylls a and b. Furthermore, it now appears that the microtubules so essential both for mitosis and for flagellar and ciliar motility may be derived from a symbiotic relationship with very primitive spirochetes.

LOCAL INTERBREEDING
POPULATION: THE UNIT OF SELECTION

By whatever means, innumerable devices were invented and assembled, leading to the vast diversification of multicellular life-styles: life-styles of habitat, of locomotion, of food types, of food gathering, of internal management, and of social behavior, including, above all, mating behavior. Beyond this evolutionary stage, the unit of natural selection in exclusively sexually propagating groups is no longer the individual but the *local population of actually interbreeding individuals*. Local populations, interconnected through slight gene flow among them, constitute a new entity: the species.

ANIMALS: COMPLEX BEHAVIOR,
CENTRAL NERVOUS SYSTEM, AND MIND

Complex social behavior, complex nervous system, and complex mind evolve together in animals. Why not in plants? Plants, too, respond to signals of various modalities, and they, too, respond with fantastic sensitivity. Plants, too, have to coordinate these signals. Plants, too, have complex mating behavior. Plants, too, have circulation of body fluids. But plants differ from animals like Germany from France: much less centralization. Germany could be dismembered and continue to thrive. France without Paris, and Paris without France are unthinkable.

Mind requires an organization with a strong center, the central nervous system (CNS), so we will confine ourselves to animals in this discussion. Crudely, we can speak of the development of the CNS of vertebrates as occurring in three stages—brainstem, midbrain, and cortex—corresponding to the regulation of vegetative functions (breathing and circulation), programs for fixed behavior patterns in response to external stimuli (midbrain), and conscious, thought-controlled behavior (cortex).

The cortex is the organ of perception, of ideation, of consciousness, of language. Four big words. Let us look at them in turn.

CORTEX: ORGAN OF PERCEPTION

Naively, you may imagine that visual perception is the image on your retina as seen by your conscious mind. Not so; many steps intervene between retina and consciousness, processing steps that abstract information from the pattern of excitation in the retina. The shifting of this pattern caused by saccadic motions and by voluntary eye movements is filtered out. When you move your eyes or head the world stays fixed even though the image on the retina wanders. This is one of numerous "perceptual constancies" brought about preconsciously by some abstracting circuitry in the brain circuitry, which evolved to permit you to orient yourself in the real, objective world. It evolved with the evolution of the brain, over millions of years; it is not a matter of learning by the individual. Other such "perceptual constancies" are the sizes of objects as they move closer or farther, the color of objects irrespective of the qualities of the illuminating light, the shapes of objects irrespective of perspective, the loudness of sounds irrespective of distance, and innumerable others. It was one of the basic misconceptions of philosophers, until quite recently, that the mind deals with "primary sense impressions" and that the *individual learns* to make the abstractions. Kant's claims of the "a-priori-ness" of such categories as space, time, and causality as necessary forms of perception almost hit the nail on the head: These categories are indeed "a priori" for the individual, but they do not fall from heaven; they are matters of evolutionary adaptation, designed for survival in the real world (Lorenz, 1959, 1962).

CORTEX: ORGAN OF IDEATION

The cortex is the organ of ideation. Let us consider the ideas of "object" and "number."

We are not born with these concepts ready at hand. They develop during the first years, through a series of stages. Piaget was the first to grasp the immensely interesting fact that the cognitive human mind is not something with which we are born but something that develops like any organ or part of the embryo; his life-long study of the development of the concepts of space, time, motion, object, causality, and number has helped provide a clearer understanding of the interplay of evolution and the experience of the individual, the interplay of the innate and the learned, that produces this development. (Refer to Flavell, 1963, for one of the best guides to the work of Piaget.) Let us look briefly at two of these concepts, the object concept and the number concept.

Object Concept

The development of the *object concept* during the first two years of an infant's life is preceded by that of eye-hand coordination. Initially, the infant has the grasping reflex for anything touching the hand but cannot grasp for something he sees. He may increase his hand waving and by trial and error reach and grasp something seen. But "something seen" is not yet an "object," is not something that has permanence whether seen or not, whether in motion or stationary. The object concept arises in the context of structuring space in the surroundings, where structuring really means coordinating and assimilating the space of the visible, the audible, the tactile, and the motor worlds. In the first stage of development, this space is only that which is immediately reachable by arms, legs, or mouth. The rest of the world remains the infinite firmament. In the next stage, memory enters, the memory of an object played with a few seconds earlier. After his attention has wandered the infant expects to find the object in the same place and returns to it. Next, you can play "hide and seek" with him, hiding an object: he will expect it to continue to exist in its hiding place and look for it and find it. In the next stage, the infant clearly forms mental images and is able to plan and to follow complex spatial transformation.

This object concept is vital for our orientation in the world, is an ancient part of our mind: The bizarre thing is that we have to abandon it in modern physics; the category of identity, of permanence, had to be eliminated in quantum mechanics. One can "identify" an electron; one can never say whether an electron "seen" now is "the same" as one seen later; particles are annihilated and created out of the vacuum. This abandonment of absolutely basic concepts, "a priori" in Kant's sense, is the hallmark of modern science, as we shall see in several examples that follow.

Number Concept

It is a long way from the concept of an object and its continued existence when out of sight, hearing, and touch, to the *concept of number*. This concept has nothing to do with the ability to recite the names of numbers. It involves the ability to see the equivalence of two sets, for example, a set of candies and a set of pennies—equivalence meaning that you can pair them off without having any left over. The difficulty of the task consists in disregarding (abstracting from) distracting features: whether the candies are spread out farther than the pennies, whether they are larger than the pennies, whether they are arranged in a line or jumbled, etc. It takes a high degree of mental operation to do these spatial rearrangements and abstractions, and the ability to do these operations (concrete mental operations) does not come until age 7-11 years. Before that stage, the child does not have the concept of cardinal number, the idea of equivalence between sets of objects. The ability is not related to cultural influences, it is an ability that develops with growth entirely as an unfolding of a genetic program.[2]

Where do these remarks about the number concept leave the issue of the "Reality of the Mathematical Universe?" Is the fact that man alone has the ability to elaborate this concept into a mathematical discipline, the fact that it develops relatively late in the child, the fact that it requires special processes of abstraction, mental powers undoubtedly rooted in cortical structures—are all these facts relevant to the question of whether the numbers were "invented" by man or whether they constitute an independent reality to be "discovered" by man, be it in evolution or in development? Consider theorems about the totality of numbers, for instance, the theorem that the series of prime numbers is infinite. The ingeniously simple proof of this beautiful theorem is found in Euclid.[3] How is such a thing possible? How can the capacity to perform a few very simple concrete mental operations permit us to discover truths extending to infinite sets? How can such "absolute" certainty be attained by a creature with a cortex evolved under selection pressures acting at prehistoric times? Hereon hinges much of the battle regarding the foundations of mathematics of the last

---

[2]Work by O. Koehler has shown that some birds (ravens, pigeons, parrots) can be trained to abstract numbers (up to 7) from a group of objects presented in an arbitrary configuration. (See Eibl-Eibesfeldt, 1970).

[3]Euclid considers the question whether the series of prime numbers (2, 3, 5, 7, 11, 13, 17 . . .), which gradually thins out, terminates in a largest prime number. He proves that it cannot terminate because the assumption of a largest prime leads to a contradiction: If a largest existed, then a larger number would be prime, namely, the product of all primes plus one.

hundred years: Cantor's attack on the paradoxical aspects of infinity; his distinction between denumerable sets and the continuum of the real numbers; Russell's and other paradoxes of set theory; Hilbert's tremendous program of "Proof Theory;" and Gödel's demonstration that the goals of this program are unattainable.[4]

The lesson of all of these powerful efforts seems to be that an understanding of the nature of mathematics cannot be attained from studies within mathematics and mathematical logic alone: We must look for it in the wider context of consciousness and language.

CORTEX: ORGAN OF CONSCIOUSNESS AND OF LANGUAGE

The cortex is the organ of *consciousness* and of *language*. Do consciousness and language necessarily go together, are they synonymous? Certainly not synonymous, otherwise we would never be at loss for words to express our thoughts. However, can we be conscious of something that would be *impossible* for us to verbalize?

Lateralization: Does the Right Hemisphere Have Consciousness?

Here the work on *lateralization* of various cortical capabilities has revolutionized our insights (see Desmedt, 1977; Sperry, 1974). Normally the two halves of the cortex are integrated so intimately that investigation is very difficult. Some years ago surgeons introduced an operation on patients with severe epilepsy. In this operation the "corpus callosum," a massive strand of fibers connecting

---

[4] Gödel showed that *any* system of axioms rich enough to encompass arithmetic with integers permits the formulation of statements the meanings of which are quite unambiguous but the truth of which is not decidable by deduction from the axioms. Gödel's method of proof does not permit one to determine for any particular conjecture whether it is decidable or not. This situation puts mathematicians in a painful state of anguish. Suppose an eager mathematician struggles for years but fails to prove or disprove the truth of Fermat's famous conjecture (that the Diophantine equation $a^n + b^n = c^n$ has no solution in integers for any $n > 2$). He may comfort himself with the thought that his inability to settle the matter is not because of his incompetence but because there is no way to decide the issue. However, he cannot feel much relief because he can *never* be certain that the matter is undecidable. Indeed certainty of undecidability, in cases of this kind, would imply certainty that no counterexample could be found and would thus prove the conjecture to be correct, i.e., it would imply a contradiction. Before Gödel mathematicians lived with the faith that any unambiguously stated mathematical conjecture must be decidable. Since Gödel this faith is shattered.

the two hemispheres, is severed. On superficial study such patients very surprisingly seem to suffer no loss at all of their normal perceptions, their motor activities, or their speech. Finer studies, however, revealed a fantastic situation: Visual or auditory or tactile inputs can be designed so as to reach only one of the hemispheres. When this is done it is found that the other half literally does not know about it. This can be tested by asking the patient (or the animal) to identify—say, by touch with his right hand, controlled by one hemisphere—an object seen in the left half of the visual field, an input leading to the other hemisphere. He cannot do it, even though the verbal instruction is given to both halves. More striking, it turns out that the right half of the brain is incapable of verbalizing what it "knows," even though this knowledge is clearly present, because the right half is able to use it for solving complex mental tasks. The dichotomy goes so far that the patient may show an emotional response, for example, smile when seeing a picture with his right half, but, if asked *why* he smiles, his verbalizing left half can only admit ignorance. These great discoveries show that we have *two* minds under one roof, two minds normally so well integrated that their separation is inapparent: they talk too much to each other (via the corpus callosum), and then talk with one voice controlled by the left half, to the outside world.

Does the right half, the nonspeaking half, then, have consciousness? Yes, it certainly has a mind in the sense that it can hear and understand speech, and rationally answer questions, not by speech, but by solving problems. Whether or not this is called consciousness is a matter of terminology, and terminology in this area is not settled at present.

What is special about speech, then, in relation to the mind?

Language: Man, Bees, Monkeys

For the purposes of this chapter, I wish to define language by four criteria, which are satisfied by human language.

1. We form a large number of *symbols* (words) by a combinatorial procedure from a small number of elements: phonemes in the case of spoken language, kinemes in the case of sign language (of the deaf).
2. We form *sentences*, bringing words into logical connection by a finite number of rules of grammar and syntax—an unlimited (denumerably infinite) number of sentences.

3. We use these sentences for *socialized actions*.
4. We have the inherited capability of *learning any of a very large number of languages* (5,000–10,000 now current) from our elders, learning words, grammar, and syntax. The phonemes as such are largely inherited and a matter of development.

Learning in general is, of course, a much older capability than learning a language, including learning from others, parents especially. Horses learn from their mothers not to shy from trains and cars, and birds have been shown to be able to learn how to pick aluminum tops from milk bottles from one who invented the trick.

If we try to apply the above four criteria to "the language of the bees"—the well-known symbolic dances they perform in their hives to transmit to their mates the direction and distance of food sources—we see that they do form symbols from a few elements. They do not form sentences with grammar and syntax. They do use the symbols for socialized actions. They do not learn the language. They are born with it. Thus, for purposes of this discussion, let us not call the bee's system of communication a language.

With respect to monkeys and apes the situation is different. Some monkeys produce a considerable number of phonemes, and they do produce a number of symbolic words (several different warning cries appropriate for different situations, threatening cries, etc.). These are used socially, but they are not used to construct sentences, and they are probably not learned from the parents. However, chimpanzees *are* capable of *learning* to communicate with their *keepers* using a variety of nonvocal models to construct simple sentences: kinemes of American Sign Language (Gardner and Gardner, 1975) or colored button recognition and handling (Premack and Premack, 1972; Rumbaugh and Gill, 1976).

Much of language concerns memory: a fluent speaker has thousands of words stored in his long-term memory, as words of *his* language, and, if he knows several languages, several words stored separately for the same object. How he accesses this memory when the word is presented is a mystery. Each word is stored with related information: its meaning, its sounding, its spelling, its grammatical category, etc.

Memory, learning, and retrieval of stored information can be studied with chess, instead of language. Chess is much more circumscribed than language; competence in chess can be measured quite well, yet it is reasonably rich and complex. How do chess masters differ from ordinary players? Do they analyze each situation more completely, in greater depth, looking ahead more moves? Astonishingly, it turns out that they do not; in fact, they explore less. Instead, they have a huge store of "chunk memory" of partial aspects of each situation. The better the master, the larger the number of chunks, up to about 40,000, like the word memory of a highly literate person, and similar to it in the capability to quickly access each chunk, and having a great deal of related information stored with each chunk. Each real chess-board situation presented to a master is immediately seen as a set of a few "chunks," like a meaningful sentence (Simon and Chase, 1974).

In addition to words, the fluent speaker stores in his memory a prodigious number of grammatical and syntactic rules of *his* language. As adults we find out how difficult this is when we try to learn a foreign language. Yet, every child of any culture has the ability to *infer* the rules of any language in a relatively short time, under conditions in which he is exposed to short time, under conditions in which he is exposed to unsystematic samples, many of them fragmentary and even faulty. This accomplishment is equivalent to learning an enormously complicated theory. Yet, even people with very low IQs have it. There is very little reinforcement needed in this process, and it happens at an age when the child is quite incapable of comparable analytical learning.

Thus, here is a capability of the human mind that, astounding as it is, has been very refractory to analysis. The attempts of the linguists even to *describe* it, in terms of language universals, transformational grammar, etc., have yielded precious little insight. Even more frustrating have been the attempts to understand the neural network underlying it. Yet, no one doubts that the basic aspects must have a clear-cut machinery.

EVOLUTION OF LANGUAGE:
DEVELOPMENT IN THE INFANT

To get at the biological meaning of language we would dearly like to know how and when it *evolved*, but here we draw an almost total blank: Is fluent language as old as the human species, now clearly documented for at least *three million* years, or did it evolve only with the domestication of fire and develop around the hearth, some *forty thousand* years ago, when there was time for leisurely talk in relative safety from predators? Opinions differ radically on the time and mode of language evolution, and data simply are not to be had.

It is suggestive, in this context, to turn to studies on *language development* in the infant, again pioneered by Piaget. The earliest function of speech in the child is very clearly *not* communication, but symbolization. The first signifiers the child uses are *private* symbols, not even verbal symbols, but rather one object serving as a symbol for another object. These lead to internalization and representation. Communication with others through words or signs comes later. At six years, more than half of the speech utterances of children in kindergarten situations are *egocentric* (not serving communication: the children carry on monologues, they "think aloud" their ongoing actions). Language is a sympton of thought structure, a dependent variable. As every parent knows, children find it difficult, even unnecessary, to impart explicit information. When forced, the child "reads aloud" his ongoing cognition, not concerned with communication. He already knows the information. This behavior contrasts fundamentally with vocalized gestures of animals, which are inherited and which have nothing to do with internalization. Thus, human language is a unique capability, of which at most only the crudest beginnings can be discerned in his closest relatives, the higher primates. Biologically, it has two functions: 1) to serve as a crutch to representational and analytical thinking (a private affair), and 2) to provide a means of communication (a social affair).

LANGUAGE: VEHICLE OF
CULTURAL AND BIOLOGICAL EVOLUTION

The combination of these two above mentioned aspects has permitted language to become *the* vehicle of cultural progress—orderly systems of knowledge, transmitted from generation to generation and, by translation, even across language barriers and, eventually, to the development of the sciences.

Ability to learn and ability to transmit learning: these two abilities are not restricted to animals with language, but within the framework of language they have evolved to their vastest expansion. It is very likely that in the wake of this process a decisive and somewhat sinister feedback loop (Bresch, 1977; see chapter appendix for translation of relevant passage) between biological and cultural evolution has occurred, running somewhat as follows:

1. The cultural evolution of language (formation of dialects) is a much faster process than any biological evolution by natural selection between rival local populations of a species.
2. Therefore, conspicuous linguistic differentiation between local populations quickly outruns biologically inherited differentiation.
3. Linguistic differentiation thus permits easy recognition of conspecifics belonging to "us" or "them," thus facilitating intergroup rivalry, internecine warfare.
4. For species on top of the food chain, i.e., those with no predators threatening to diminish their number, to proliferate quickly is not a selective advantage but, on the contrary, a disadvantage.
5. Natural (biological) selection between these groups works in favor of those best equipped, language-wise, and fiercest in aggressive instincts directed against conspecifics that have a different "culture," that is, behavior patterns and especially *cultural* behavior patterns, language.

It is a fact that in many "primitive" tribes the joy of killing individuals of the tribe in the next valley surpasses almost any other joy, and it seems at least very plausible that the high selective premium associated with this form of aggression has been responsible for the "success" of *Homo sapiens*, and for the extinction of many sidelines in the evolution of the species. The instances of near or complete extinction of many "native" populations in historical times probably are examples of what happened throughout Eurasia and Africa for as long as "culture," i.e., transmission of learned language, has been around.

THE SCIENCES: TRANSCENDENCE OF FORMAL OVER CONCRETE OPERATIONS

In developing the sciences, what has happened to the old concepts of object, number, time, topological space, projective space, metric space, causality? What has happened to all the concepts that together constitute our naive view of physical reality? What has happened to this evolutionary acquisition of immense adaptive value, enabling us (as higher vertebrates) to get along in "the world," so conceived? The funny thing is that science has pulled the rug out from under this structure. *Special relativity theory* has replaced the space-time frame with an abstract one in which a man may go on a trip and return a younger man than his twin brother, a claim that is irreconcilable with our concrete mental operations of space and time. *General relativity theory* tells us of "singularities in space," black holes with an "event horizon" from which no signals can emerge, of finite but not bounded space, concepts that we can learn to manipulate in a formal way but that our minds cannot "visualize." *Quantum theory*, the worst offender, does away with object identity, trajectory of objects (electrons do not run around in orbits); it proclaims a conspiracy of Nature that forces us to *choose*, to make "either-or" decisions.between various aspects of reality that exclude each other in any observational act. Is this a Kierkegaardian notion in that every observational act becomes an existential one? Have the physicists become religious thinkers?

Indeed not, because the "choice" is not an ethical one; it is not even the choice of an individual observer, but one that concerns collective observations on, for example, light quanta: pass them through a contraption that permits statements about their *circular* polarization (right or left circular), or through a contraption that permits statements about their *linear* polarization (vertical or horizontal). "We" make this choice, and these choices materially exclude each other, because any one quantum, once it is observed, i.e., recorded by a counter, is irreversibly gone. Such is the individuality, the quantum nature, of any atomic interaction that is involved in constructing an object world. It always leaves this object world with a residue of uncertainty (Heisenberg's principle) and limits predictions to statistical ones.

THE CARTESIAN CUT

This bizarre dialectical situation goes to the heart of the concept of the "reality of the physical world," so basic to the evolution of our mind from that of the mosquito.

> O Seligkeit der *kleinen* Kreatur, die immer bleibt im Schoosse der sie austrug.
> O Glück der Mücke die noch innen hüpft, selbst wenn sie Hochzeit hat . . .
> (O bliss of tiny creatures that *remain* forever in the womb that brought

them forth! Joy of the gnat that still can leap within, even on its
wedding day ...)
Wer hat uns umgedreht, dass wir, was wir auch tun, in jener Haltung
sind von einem welcher fortgeht?
(Who's turned us round like this, so that we always, do what we may,
retain the attitudes of someone who's departing?)

(Rilke, 1939)

For millions of years we have been animals who know the dicho-
tomies: actor/onlooker, I and the world, mind versus real world, a
confrontation between an inner world of thoughts, volitions,
emotions, and an external world. (Jaynes' (1976) book, referred to
earlier, defends the thesis that the dichotomy is of very recent
origin—3,000 years.)

This is the Cartesian cut, into a *res cogitans* (the soul) and a
res extensa, two distinct substances, which has been the stance of
science for 300 years and has been the blight of psychologists,
whose job it is to cope with both sides and to tie them together in
some fashion. Is the tree I see in front of me the same as the object, or are the two things distinct? When we think of the retinal
image of the tree—the conduction and processing of this
image occurring in the neural network of the retina itself, in
the lateral geniculate nucleus, the striate cortex, visual cortices II
and III, etc., and finally emerging into consciousness—we must
realize that what consciousness sees is literally worlds apart from
the object. On the other hand, does it make sense to take the object
and its perception apart in this way? There is only one reality: the
act of seeing, the seeing of what our language forces us to call "an
object." These are the opposing points of view of the dualist and
the monist. Contingents of philosophers have straddled the barri-
cades between them in assorted postures.

HOW "OBJECTIVE" ARE THE
OBJECTS OF THE NATURAL SCIENCES?[5]

The notion of the "objective physical reality of the physical world,
independent of the observer" is a notion we form in earliest infancy,
and we form it with the aid of mental equipment evolved over
millions of years of adaptive evolution. It is a notion that has been
most necessary for survival, for getting along in the cave, and even

---

[5]This passage (pp. 161–163) is adapted from a lecture by N. Bischof (1969).

before the cave, in the trees! It is also the notion that has been most solidified, hardened, codified by the development of the classical physical sciences themselves, into the "physical law." It has been said, and it is a common belief, that these physical laws describe the external world in an *objective way* and that they reduce this description to *numerical relations*. Let us look at these claims more closely, for instance, at the relation $S = \frac{1}{2} gt^2$. It is indeed a relation between numbers. However, to make it a law of physics we need to know that S is a distance, t a duration, g an acceleration. The numerical relation as such does not express a physical law. The law is expressed only if we understand that the numbers are measures of a quality of the thing measured: a spatial length represents something completely different in quality from a duration or an acceleration. In addition to the qualities of the quantities measured, we need to know where and what to measure. We need to know the class of actual situations to which the law refers. In fact, the physical law, far from hanging out totally detached from the observed object, refers very explicitly to situations actually or potentially experienced by an observer and to nothing else. This is true, it may be added parenthetically, even if we make statements about the origin of the universe, e.g., the Big Bang, even though, by no stretch of the imagination could there have been an observer around to observe it. It would be an illusion to think that physical laws describe the external world independent of the observer.

In what sense, then, is physics objective? We say it is objective in being "reproducible" for each observer and "the same" for different observers. These two criteria are the pride of the physical sciences, and they are indeed met. Information stored in enormous handbooks of physics and chemistry and the rigid core of the textbooks of these disciplines give vivid testimony to the objectivity. One might characterize the physical sciences as those for which explicit connection to actual experience constitutes an annoying constraint from which they are trying to liberate themselves more and more. The "physical law" is supposed to refer to larger and larger classes of experiences. The infant's first construction of space-time frames, of notions of persistent objects, and of causal connections constitute giant steps in this direction. Nevertheless, the fact remains that the physical sciences represent the actual or potential experiences and observations of individuals, in however abstracted a form, and, as such, are as psychic as any emotion or sensation. Both the blue of a summer sky and the 4,400 Å of the wavelength of its light refer to experiential acts, differing principally in the affective components accompanying these acts and their expressions. The statement is often made that "blue" is a private sensation that cannot be identified with another person's sensation because it is subjective. The same is true for the size of a table. How do I know how large you see it? Indeed, your impression must be different because you are farther away. So we measure it with a ruler. But measurements of length, i.e., noting coincidences of marks on the ruler with edges on the table, are private acts, too, comparable only to the extent that we have linguistic expressions for them.

It is the parsimony of the number of elements singled out that makes the duality of observer and observed so successful and gives the illusion, in physics, that the object is totally distinct from the observer. This distinctness is true in the sense that we do not mention the observer when we say, for instance, that the present temperature of the black-body radiation left over from the Big Bang at the beginning of time is $3°$ K, or that a supernova exploded 10^9 years ago. Any such statement is linguistic in the first place and, as such, meaningful only within the framework of the total scientific discourse, reflecting individual and collective experiences and acts.

PHYSICAL VERSUS MENTAL: NOT A BASIC DISTINCTION

My point is this: the distinction between psychic and physical is not at all a radical one but a matter of degree. Consider eye-motor coordination when you try to hit a tennis ball: your eye sees it coming, your brain commands the arms and legs to make appropriate motions, the eye and the proprioceptor apparatus continue to monitor and correct this output by comparing it with a mental image of what should happen, a classical feedback loop containing the conscious visual picture, surely a psychic element, as part of the interactive network. It is true that meter readings, etc., lend themselves to quantification more easily than mental images of an intended motion, and for that reason they are communicated more easily both in informal and in mathematical language. In principle, however, both are psychic and both are physical.

I therefore claim that the antithesis of external and internal reality is an illusion, and that, in fact, there is only one reality. Quantum mechanics has simply served to remind us of this fact, which seemed to get lost in the abstractions of physical science.

The issue of external versus internal reality looks especially confusing when we consider more closely those situations in which the object is the consciousness of ourselves or of someone else. Say, you are thirsty. In what sense is your thirst a physical quantity? Whatever "thirst" may be, it is something you can have more or less of. In principle, therefore, it is measurable, either by behavioral tests or by physiological correlates. While it is useful to make the practical distinction between psychic terms and nonpsychic terms, the distinction is not more basic than that between liquids and solids.

CONCLUSION

In this long (much too long and much too short) chapter, I have attempted to look at the oldest problems of epistemology: what is *Truth* and what is *Reality*? I have tried to respond to the raised eyebrow, as it were, implied in the title of this chapter: "Mind from Matter??" We start with the naive question: How can mind emerge out of dead matter in the course of "purely physical" processes? Mind, which then looks back upon itself and says, "Aha, this is how I came about" (like Baron Münchhausen, pulling himself by his hair out of the mud). I have avoided the language of the "emergent evolution" school (Morgan, 1927), a school that in my opinion has not made a constructive contribution to these questions; on the contrary, it has camouflaged them behind an appealing metaphor. I have looked at the questions as a scientist, as a student of evolution: evolution in the widest sense—evolution of the Universe, evolution of Life, evolution of Mind, evolution of Culture.

The point of view of the evolutionist has the principle virtue of *forcing* us to view Mind in the context of other aspects of evolution: Parallels to other forms of adaptation (organs of locomotion, of sensing, cross-modal integration in the CNS, etc.) become imperative. Mind, the mind of the human adult, the object of so many centuries of philosophical studies, ceases to be an absolute. It is seen to be a product of adaptation in response to selection pressures, as is anything else in biology.

The sciences have played a peculiar role. The *classical* natural sciences served, as it were, to solidify the feeling that the adult human mind is an absolute: this mind grasps absolute physical laws concerned with absolute matter embedded in absolute space and time. The Cartesian cut between Mind and Matter is the rock on which it stands.

Modern science has gone the opposite way: it has forced us to abandon absolute space and time, determinism, the absolute object. It has shown that these "naive" notions must be replaced by more abstract formal schemes, that the naive notions are applicable only in the "middle dimensions" of space and time and energy. As soon as we go to extreme dimensions, our intuitions, i.e., our concrete mental operations, become inadequate. This is the point, exactly, at which evolutionary thinking is decisively helpful: it suggests, indeed it *demands*, that our concrete mental operations are adaptations to the life-style in which we had to compete for survival a long, long time before science. As such, we are saddled with them, as we are with our organs of locomotion and our eyes and ears. But, in science, we can transcend them, as electronics transcends our sense organs.

Why, then, do the *formal* operations of the mind carry us so much further? Were these abilities not also matters of biological evolution? If they, too, evolved to let us get along in the cave, how can it be that they permit us to obtain deep insights into cosmology, elementary particles, molecular genetics, number theory? To this question I have no answer.

Indeed, the approach I have sketched by no means resolves all puzzles, nor does it produce a grand synthesis between the Universes of Discourse now current in the various sciences. Least of all does it give a basis for a new setting of values in ethics, values relevant to the life of the individual or to social organization.

The feeling of absurdity that attaches to the notion "Mind from Matter" is perhaps of a similar nature to the feeling of absurdity we have learned to cope with when we permit relativity to reorganize time and space and quantum theory to reconcile waves and corpuscles. If so, then there may yet be hope for developing a formal approach permitting a Grand Synthesis.

REFERENCES

Bischof, N. 1969. Hat Kybernetik etwas mit Psycholgie zu tun? Psychologische Rundschau 20:237–256.

Bresch, C. 1977. Zwischenstufe Leben. Evolution ohne Ziel? R. Piper and Company, Munich and Zurich.

Desmedt, J. E. (ed.). 1977. Language and Hemispheric Specialization in Man: Cerebral Event-Related Potentials. Karger, Basel.

Dickerson, R. E., Timkovich, R., and Almassy, R. J. 1976. The cytochrome fold and the evolution of bacterial energy metabolism. J. Mol. Biol. 100:473–497.

Eibel-Eibesfeldt, I. 1970. Ethology. Holt, Rinehart and Winston, New York.
Flavell, J. H. 1963. The Developmental Psychology of Jean Piaget. Van Nostrand Co., New York.
Gardner, R. A., and Gardner, B. T. 1975. Early signs of language in child and chimpanzee. Science 187:752–753.
Hong, H. V., and Hong, E. H. (eds. and trans.). 1975. Sören Kierkegaard's Journals and Papers, Vol. 3, Natural Sciences section. Indiana University Press, Bloomington.
Jaynes, J. 1976. The Origin of Consciousness in the Breakdown of the Bicameral Mind. Houghton Mifflin Co., Boston.
Lorenz, K. 1941. Kant's Lehre vom Apriorischen im Lichte gegenwärtiger Biologie, Blätter für deutsche Philosophie 15:94–125. (Eng. trans. by Charlotte Ghurye in: Bertalanffy and Rapaport (eds.). 1962. General Systems VII, Soc. for General Systems Research, pp. 37–56. Ann Arbor, Mich.)
Lorenz, K. 1959. Gestaltwahrenehmung als Quelle Wissenschaftlieher Erkenntnis. Z. Exp. Angew. Psychol. 6:118–165. (Eng. trans. as above, pp. 1–36.)
Moorbath, S., O'Nions, R. K., and Pankhurst, R. J. 1973. Early archaean age for the isua iron formation, West Greenland. Nature 245:138–139.
Morgan, L. 1927. Emergent Evolution. Gifford Lectures Delivered at the University of St. Andrews in 1922. Henry Holt and Co., New York.
Orgel, L. E. 1973. The Origins of Life, Molecular and Natural Selection. Wiley, New York.
Premack, H. J., and Premack, D. 1972. Teaching language to an ape. Sci. Am. 227:92–99.
Rilke, R. M. 1939. Duino Elegies. No. VIII, 52–55, 70–73. (Trans. by J. B. Leishman and Stephen Spender.) Norton and Co., New York.
Rumbaugh, D. M., and Gill, T. V. 1976. Language and the acquisition of language-type skills by the chimpanzee. Ann. N.Y. Acad. Sci. 270:90–123.
Rumbaugh, D. M., and Gill, T. V. 1976. The mastery of language-type skills by the chimpanzee. Ann. N.Y. Acad. Sci. 280:562–578.
Schopf, J. W., and Barghoorn, E. S. 1967. Alga-like fossils from the Early Precambrian of South Africa. Science 156:508–512.
Schroedinger, E. 1945. What Is Life? The Physical Aspects of the Living Cell. Based on a Course of Public Lectures delivered under the auspices of the Institute of Advanced Studies at Trinity College, Dublin, Ireland. Cambridge University Press, Cambridge, Eng.; The Macmillan Co., New York.
Shklovsky, I. S., and Sagan, C. 1966. Intelligent Life in the Universe. Holden-Day, Inc., San Francisco. p. 210.
Simon, H. A., and Chase, W. G., 1974. Skill in chess. Am. Sci. 61:794–798.
Sperry, R. W. 1974. Lateralization in the surgically separated hemispheres. In F. O. Schmitt and F. G. Worden (eds.), The Neurosciences: Third Symposium. The MIT Press, Cambridge, Mass.

APPENDIX

The following translation of the passage from Bresch (1977) referred to in this chapter is appended here because this interesting book is not yet available in English. (Bresch, C. 1977. Zwischenstufe Leben. Evolution ohne Ziel? R. Piper and Company, München, Zürich. pp. 195–202. Translation by Jonathan Delbrück. Reprinted by permission.)

> Sometime, two to five million years ago, hominid apes became men. Where lies the distinction?
> What characteristic separates Man and animal?
> What is human about humanity?
> The premeditated killing of other members of the same species?
> The upright walk?
> Tools?
> Language?
> The change is no sudden leap. None of Man's special characteristics appeared overnight. None of them alone was responsible for his rapid further development. However, it is difficult to deny that men acted as their own most effective tool of selection, by the extermination of inferior groups, which led to rapid further development of their brains. We shall see how this came about.
> Throughout the animal kingdom we see fighting within species. In nearly all cases, the weaker one retreats, and his flight ends the altercation. The higher species have evolved, besides, a submission gesture, which immediately turns off the aggressive mood of the victor, so that he spares the loser any further punishment. The result is that in the animal kingdom, killing within a species only happens as an accident.
> It must be added here that life and death struggles within a species are not an exclusive characteristic of men. For example, lions who have taken the females from an aging rival may kill their progeny. Eagles usually lay two eggs, but only raise one offspring, which kills the other in a life or death struggle which the parents may even watch, without interfering. Bees kill every alien bee in their hive, which they recognize by smell. How is this behavior to be explained? Isn't genocide harmful to the prosperity of a species, and thereby contradictory to the basic principles of selection otherwise observed? To clear up this apparent contradiction, we must first back up a little.
> Let us imagine any well established animal species. It will eat plants or animals, and will also be food for some other animals. Natural selection has provided a genetic predisposition against genocide. This genetic information is always subject to changes, most with negative results. There will, therefore, be occasional mutants that direct their aggressions at members of their own species. These "defect-mutations" are for the time being limited to a certain region (of their origin). If we compare the number of progeny in this 'killer-region' with the remaining "peaceful" area, we can see that the aggressive population

reproduces less efficiently than the peaceful one, which is threatened only by enemies, and not from within. The smaller numbers of surviving progeny mean their gradual disappearance. The peace lovers have a selective advantage.

Therefore, such aggressive mutations may appear in any species, but have only a transitory existence.

Now there are some exceptional ones, the "kings" of the animals, who may eat other animals, but, with their well-protected upbringing, never in their lives are vulnerable to other predators. For these species at the top of the food chain, different rules of selection apply. A high reproductive rate is no longer advantageous to animals with such exceptional capabilities. The only remaining threat is of starvation, which is inevitable if they overpopulate an area. With the near elimination of their prey, their own numbers fall to nearly zero. Rapid reproduction is therefore only an advantage when an outside enemy keeps the population density in balance.

So, when animal species at the top of the food chain encounter mutations which reduce their reproductive rate (and such "defectives" will appear again and again), it is not to their disadvantage, as long as sufficient offspring are produced for the continuation of the species. Indeed, a low population density avoids famine and the associated risk of decimation to *extremely* low numbers.

In addition, a smaller number of progeny can be more effectively raised by the parents, which is important for the transfer of information to the next generation. For all these reasons we can expect populations even at the top of the food chain to stabilize at a constant density, through mutations that bring their reproduction into balance.

There are many ways to throttle the rate of propagation, for example the delayed onset of sexual maturity, which takes 15 years for a human being, while another mammal, the mouse, is already mature in nine weeks. The mother mouse can also be fertilized immediately after giving birth, while the chimpanzee is sterile for years afterward. The gestation period of the mouse takes just three weeks, and she will have three to 10 per litter, while a cow elephant produces in her lifetime only about four offspring, with a 21-month gestation period.

A completely different solution to the overpopulation problem is killing within the species. Only in exceptional species can such a defect become well established. Even in those cases, the killing must be limited to certain situations, or to special selective criteria. A mutant population which indiscriminately goes about killing each other off will soon extinguish itself. Limited killing is, however, tolerable and may serve, in conjunction with other reproductive characteristics (time to puberty, gestation, litter size, delayed fertility, etc.), to establish a fairly constant population density.

Once a limited form of genocide is established, it becomes itself a powerful tool of selection, as consistently the stronger, quicker, more intelligent animals survive. Their superiority over other species is thereby strengthened. Originally a regressive mutation, like the loss of pigment in burrowing animals, the killer-mutation leads to further evolution.

Humans belonged to this small class of exceptional species. Their tribes had practically no enemies to fear, and so, liberated by his intelligence from the threats of other animals, man became his own enemy. This development was facilitated by various factors, founded in man's new capabilities:

Weapons—even if they were simply stones thrown—killed much quicker than claws and teeth. Also, they were effective over a distance. There was therefore no longer any use for mutually understood submission gestures, and their associated limitations on aggression.

In addition, there was the following important aspect: With expanding capabilities of the brain, the animals' reactions developed from automatic responses to more complex patterns of instinctive behavior. These in turn retreated into the background, as the individual's personal experiences gained importance in decision making. When finally linguistic communication made it possible to establish a group tradition, the instinctive reaction lost all its value compared with the flexibility of reactions that could be calculated by the brain from a combination of group traditions, individual experiences, and the particulars of any given situation. The weaker the genetic predisposition, and the greater the freedom of choice of the brain, the better the group's chances for survival. The instinctual drives behind human behavior were thus being reduced and replaced by intellectual control through "custom." This tendency is the basis of becoming human, even though man today occasionally still acts on the impulse of primordial drives.

Particularly in the area of threat and submission gestures, language was supplementing gesture and mimicry. Within the tribe, that was understood, but one who could not even speak, one who merely emitted meaningless grunts, was simply not human. With humans one could speak, and come to an understanding. One who spoke differently could not beg for mercy. Language was the distinguishing factor in a much narrower social group. Anybody could look human—language made the difference.

Presumably just this new differentiation led to a criterion of whom one could kill. The strangers no longer belonged to the same kind. Common language produced a completely new and intense feeling of belonging together. The closer the inner cohesiveness of the group, the more easily everyone else is labeled as enemy.

Index

Actualization, of the individual, 104–105
Adaptation
  of biological systems, 3–4
  of organisms, in higher species, 2
Adaptive behavior, types of, 147–148
Adenosine diphosphate (ADP), in ATP synthesis, 78
Adenosine monophosphate (AMP), in ATP synthesis, 78
Adenosine triphosphate (ATP), 149
  synthesis of, 78
Adenosine triphosphatase (ATPase), 78
Adenylates, preformed radioactive aminoacyl, interactions of, 43–44
ADP, see Adenosine diphosphate
Aging, evidenced by the formation of junctions, in proteinoid microsystems, 67
Ala, see Alanine
Alanine, 32, 34
Algae
  blue-green
    ancestral traces of, 4
    establishing on Mars to make planet habitable, 18
    thermophilic, 38
  symbiotic relationship with fungi, 10–11
American dream, and the web of interdependence, 95
Amino acids
  characteristics of, 39
  coded relationship between oligonucleotides and, 42
  origin of, 26
  polymerization of, presence of conditions for, 28
  role of, in Life's origin, 4, 5
  selectively reactive features of, 37
  self-instructing, 34

self-ordering of, 27, 79, 80
  and the chicken-egg questions, 28–34, 36
  wider significance of, 37–38
  thermal polycondensation of, forces operative in, 35–36
α-Amino acids, thermal copolycondensation of, 38
Ammonia, in early atmosphere, 4, 6
AMP, see Adenosine monophosphate
Ampholytic macromolecules, binding ability of, 45, 48
Anticodonicity, 43, 44
"Aperiodic crystal," 143, 144
Apollo program, 26
Associations, see Biological associations; Creative associations; Symbiotic associations
Atmosphere, composition of, before advent of Life, 4, 6
ATP, see Adenosine triphosphate
ATPase, see Adenosine triphosphatase

Bacteria
  behavior of, 147–148
  morphology of, compared to proteinoid microsystems, 55
  sensory system of, 115–117
Becoming, principle of, 99
Bees, "language" of, 156
Behavior
  principles of organization of, general nature of, 115
  see also Adaptive behavior; Social behavior
Behavior of systems, description of, in the discipline of physics, 142
Bidirectionalism, between amino acids and nucleotides, 35, 43, 79, 83